Carbohydrate Synthons in Natural Products Chemistry

ACS SYMPOSIUM SERIES **841**

Carbohydrate Synthons in Natural Products Chemistry

Synthesis, Functionalization, and Applications

Zbigniew J. Witczak, Editor
Wilkes University

Kuniaki Tatsuta, Editor
Waseda University

American Chemical Society, Washington, DC

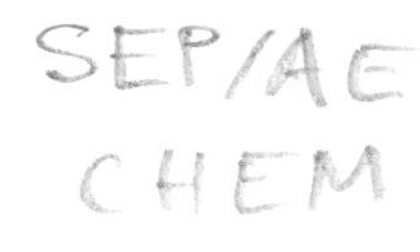

Library of Congress Cataloging-in-Publication Data

Carbohydrate synthons in natural products chemistry : synthesis, functionalization, and applications / Zbigniew J. Witczak, editor, Kuniaki Tatsuta, editor.

p. cm.—(ACS symposium series ; 841)

"218th National Meeting of the American Chemical Society, San Francisco, California, March 26–30, 2000."

Includes bibliographical references and indexes.

ISBN 0–8412–3740–9

1. Carbohydrates—Congresses. 2. Organic compounds—Synthesis—Congresses. 3. Chirality—Congresses.

I. Witczak, Zbigniew J., 1947- II. Tatsuta, Kuniaki, 1940- III. American Chemical Society. Division of Carbohydrate Chemistry. IV. American Chemical Society. Meeting. (218th : 2000 : San Francisco, Calif.). V. Series.

QD320 .C39 2002
547′.78—dc21 2002028371

The paper used in this publication meets the minimum requirements of American National Standard for Information Sciences—Permanence of Paper for Printed Library Materials, ANSI Z39.48–1984.

Distributed by Oxford University Press

ac

PRINTED IN THE UNITED STATES OF AMERICA

Foreword

The ACS Symposium Series was first published in 1974 to provide a mechanism for publishing symposia quickly in book form. The purpose of the series is to publish timely, comprehensive books developed from ACS sponsored symposia based on current scientific research. Occasionally, books are developed from symposia sponsored by other organizations when the topic is of keen interest to the chemistry audience.

Before agreeing to publish a book, the proposed table of contents is reviewed for appropriate and comprehensive coverage and for interest to the audience. Some papers may be excluded to better focus the book; others may be added to provide comprehensiveness. When appropriate, overview or introductory chapters are added. Drafts of chapters are peer-reviewed prior to final acceptance or rejection, and manuscripts are prepared in camera-ready format.

As a rule, only original research papers and original review papers are included in the volumes. Verbatim reproductions of previously published papers are not accepted.

ACS Books Department

Contents

Indexes

Preface

The synthesis of new chiral organic compounds and the improved synthesis of known substances will always be a major task for the professional chemist. When constructing target molecules with multiple chirality centers, a scientist must consider either total synthesis step by step or assembly from smaller chiral blocks as an alternative approach.

Carbohydrates represent a unique family of polyfunctional compounds, which can be chemically or enzymatically manipulated in a multitude of ways. Carbohydrates have been extensively used as starting materials in enantioselective synthesis of many, complex natural products with multiple chirality centers. Synthetic organic chemistry that utilizes these carbohydrate building blocks continues to spawn revolutionary discoveries in medicinal chemistry, pharmacology, molecular biology, glycobiology, and medicine simply by providing not only the raw material but also the mechanistic insight of modem molecular sciences. This interdisciplinary approach to modem discoveries and many further innovations continue to drive the core of synthetic carbohydrate chemistry. The environmentally and ecologically friendly nature of carbohydrates is also a cornerstone in their future developments in the polymer and pharmaceutical industries and in the area of carbohydrate therapeutics in particular.

Corey (E.J Corey *Pure Appl Chem* **1969,** *14,* 30) introduced the term *synthon* in 1969 when he published his innovative strategies for the construction of complex molecules by considering a retrosynthetic analysis. Later on, Hanessian's (*Total Synthesis of Natural Products: The 'Chiron' Approach;* Pergamon Press, 1983) introduction in 1983 of the term *Chiron* reférring to chiral synthons became the general strategy of carbohydrate like symmetry in new molecular targets of many natural products.

Despite the greater awareness of carbohydrate synthons in recent years, the full potential of the carbohydrate chiral pool is still not fully used. Thus, this fact gives enormous rationale in organizing our symposium and presenting new developments by a team of world-class scientists. Consequently, publishing this symposium proceeding will assist the carbohydrate community in keeping abreast of new innovations. We hope that these few, important forward-looking topics of brand new developments from world-class leading laboratories will effectively fill the gap of previously unavailable practical information regarding the unlimited possibility of applying carbohydrate building blocks.

Among a few often-used carbohydrate building blocks, L-arabinose is one of the most important and easily commercially available monosaccharides. Next, in terms of availability and potential functionality are naturally protected 1,6-anhydrosugars derivatives such as levoglucosan and levoglucosenone. Both compounds possess enormous potential for becoming new stars among industrial chemicals, simply because of their multiple usage in many areas of industry (including polymer chemistry, biotechnology, pharmaceutical intermediates, and carbohydrate scaffolds for combinatorial chemistry approaches). Industrial production of these convenient chiral building blocks from waste cellulosic material, such as newsprint or any waste paper, could solve some environmental problems and could be classified as green chemistry. The raw carbohydrate material for the functionalization into useful building blocks must be economically feasible and cost effective; waste cellulosic materials fit into that category very well. Particularly valuable building blocks such as levoglucosenone, isolevoglucosenone, L-arabinose, parasorbic acid, dihydropyranones, 3-hydroxy-γ-butyro-lactones, 1-thio-1,2-*O*-isopropylidene acetals, ω-bromo-α-β unsaturated aldonolactones, bicyclic furanones, arabinonic acid γ-lac-one and glycosyl isocyanides are explored for their synthetic applicability in many synthetic targets of natural products of medicinal interest.

Most of the chapters in this book were presented in the special symposium *Chemistry for the 21st Century* at the 218th ACS National Meeting in San Francisco, California on March 26–30, 2000. Other chapters, not presented at the symposium, are contributions from leading scientists in the field of carbohydrate chemistry.

Most importantly, these topics will help steer the future of new developments in this area and will help promote the enormous potential of many innovations among almost all chemical industries in the new millennium. This is not a simple goal, but a 2lst century challenge to educate the industrial leaders, public, and governmental funding agencies about the enormous potential and usefulness of these traditional and new carbohydrate synthons as chemicals for the 21st century.

Acknowledgment

We thank all the authors for their excellent contributions to this volume. We also thank the peer reviewers of the chapters for their expertise and enormous efforts to improve the quality of the manuscripts. We are grateful to the ACS Division of Carbohydrate Chemistry for sponsoring the symposium upon which this book is based. We also acknowledge Kelly Dennis and Stacy VanDerWall in acquisitions and Margaret Brown in editing/production of the ACS Books Department for their help in coordinating and producing the book.

We dedicate this book to our wives Wanda and Yoko.

Zbigniew J. Witczak
Department of Pharmaceutical Sciences
Nesbitt School of Pharmacy.
Wilkes University
Wilkes-Barre, PA 18766

Kuniaki Tatsuta
Graduate School of Science and Engineering
Waseda University
Shinjuku
Tokyo 169–8555, Japan

Carbohydrate Synthons in Natural Products Chemistry

Chapter 1

Chiral Carbohydrate Building Blocks with a New Perspective: Revisited

Zbigniew J. Witczak

Department of Pharmaceutical Sciences, School of Pharmacy, Wilkes University, Wilkes-Barre, PA 18766

The chiral bicyclic enones, levoglucosenone, isolevoglucosenone, and new functionalized L-arabinose enone possess excellent reactivity and functionality. Their properties and application as convenient precursors in the synthesis of many attractive templates or intermediates of complex natural products are reviewed. These compounds are attracting increasing interest due to their structural rigidity and ability for stereoselective functionalization without protection, deprotection sequences necessary in many synthetic organic methodologies.

Historical Background

Carbohydrates have been extensively used as chiral starting materials in enantioselective synthesis because of their availability as inexpensive derivatives. One of the first and foremost carbohydrate precursor employed was functionalized glucose. However, glucose often does not resemble the final target and therefore requires multistep processes of functional group conversion through protection/deprotection. Alternative strategies of using functionalized carbohydrate derivatives and converting them into useful chiral building blocks offer a more efficient approach to synthetic problems. There are a few readily available building blocks that can be prepared inexpensively in a few steps, in pure form and without costly reagents.

Examples of these convenient chiral building blocks reviewed in this chapter include levoglucosenone, isolevoglucosenone and L-arabinose derivatives.

Levoglucosenone

Levoglucosenone *(1)* is an attractive chiral carbohydrate building block that can be conveniently produced by the pyrolysis ofcellulose-composed materials. Despite the low yield and the amount of solid cellulosic material necessary for pyrolysis, the efficiency and the economy of the pyrolysis process makes it an effective method. In addition, pyrolysis reduces the amount of waste cellulosic material, which is beneficial to the environment. Although levoglucosenone has been known and used for over 30 years *(2)*, it continues to have only limited applications in organic synthesis. This can be attributed to the rather conservative opinion regarding its process, purification and stability. This simple and small bicyclic enone molecule is an important and efficient chiral starting material for the synthesis of many analogs of complex natural products and its chemistry has been reviewed extensively (*1*). Only recently published new developments will be reviewed in this chapter.

During initial stages of the cellulose pyrolysis, the formation of levoglucosan can be further dehydrated by the removal of two molecules of water with the predominant formation of levoglucosenone as one of the major products. Two other products present in the complex mixture of volatile molecules are hydroxymethylfurfural and levulinic acid. The primary factors determining the preferential double dehydration of intermediate levoglucosan are probably steric factors and the overall influence of the 1,6-anhydro-ring system in the $^{1}C_{4}$ chair conformation of the pyranose. Additionally some evidence of significant differences in reactivity of *axial* and *equatorial* hydroxyls at C-2, C-3, and C-4 of the 1,6-anhydro ring as reported in the literature (*3*) likely play a significant role in the preferential elimination of water molecule from C-3 and C-4 versus from C-2 and C-3.

All the research to date supports the preferential formation of levoglucosenone despite the possibility of double dehydration with alternative formation of isolevoglucosenone. This formation has never been detected in the pyrolysate and is only available through a total synthesis.

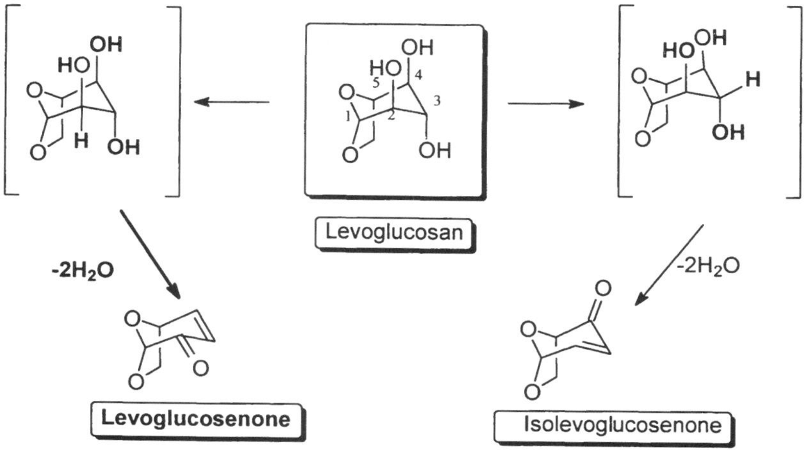

Scheme 1.

Despite the efforts of various laboratories *(1,3-10)* to promote the chemistry of levoglucosenone, isolevoglucosenone and its new analogs, applications of these remarkable materials in industry remain low. We hope that further awareness of the potential of levoglucosenone will make it a commodity product, a status that should have been granted to this molecule long ago. Thus, the goal of this chapter is to highlight all the possibilities of high potential of the carbohydrate chiral pool and put on the map all the valuable chiral building blocks, which are still little exploited.

New Chiral Building Blocks from Levoglucosenone

Among the new developments in the chemistry of levoglucosenone is the ability to functionalize the compound's C-3 and C-2 positions. These positions are very important in order to facilitate the further reactions leading to compounds with practical utility. The functionalization of the keto function by the epoxidation, using the Corey reagent (dimethylsulfoniumethylide in DMSO and THF), as reported by Gelas and Gelas *(11 -12)* is illustrated in scheme 2. The C-2 epoxide has potential synthetic utility as a precursor for highly functionalized analogs with amino, fluoro, thio, or methylene functional groups.

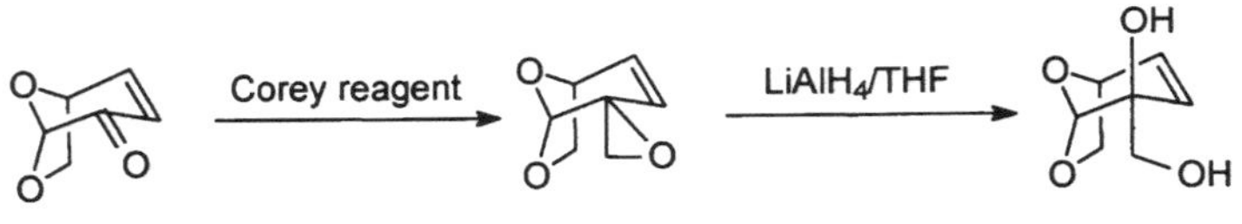

Scheme 2.

The C-3 position of levoglucosenone is strategically important and can be functionalized with thio, amino, and acetamido groups. This is usually accomplished through construction of a specific precursor bearing a good leaving group, such as iodine, bromine or fluorine at the C-3 position. New representative example of such precursors is the 3-iodo analogs of saturated levoglucosenone was synthesized in our laboratory, *(13)* according to the general methodology of selenium dioxide mediated α-iodination *(14),* as depicted in scheme 3.

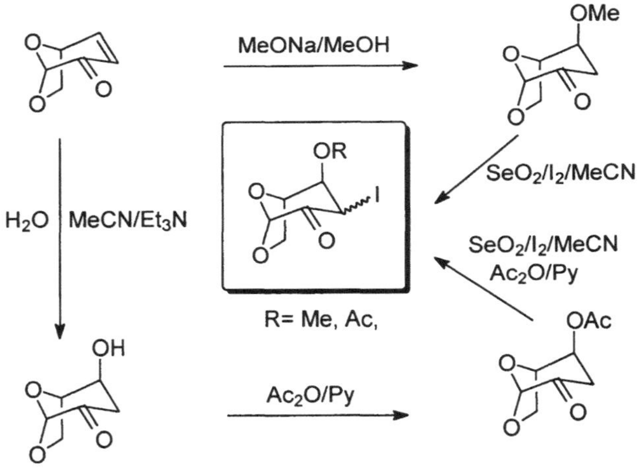

Scheme 3.

Coupling 3-iodo derivative with reactive 1-thiosugars proceeds with good yield, without inversion of configuration, and with expected stereoselectivity at C-3. This approach as depicted in scheme 4 constitutes a general methodology and opens a new route to new family of (1-3)-S-thiodisaccharides, which are otherwise difficult to synthesize under normal conditions of multistep techniques of protection/coupling/deprotection sequences. Stereoselective reduction of the C-2 keto function with L-Selectride in anhydrous THF solution produces *gluco* epimer in high (79%) yield. Conventional acetolysis in order to cleave the 1,6-anhydro ring was performed with boron trifluoride etherate in acetic anhydride solution to produce crystalline octaacetate. The final deprotection of octaacetate was carried out with an aqueous methanolic solution of triethylamine at room temperature for 8 h results in the new thiodisaccharide 3-S- (β-D-glucopyranosyl-3-thio-D-allopyranose in 89% yield.

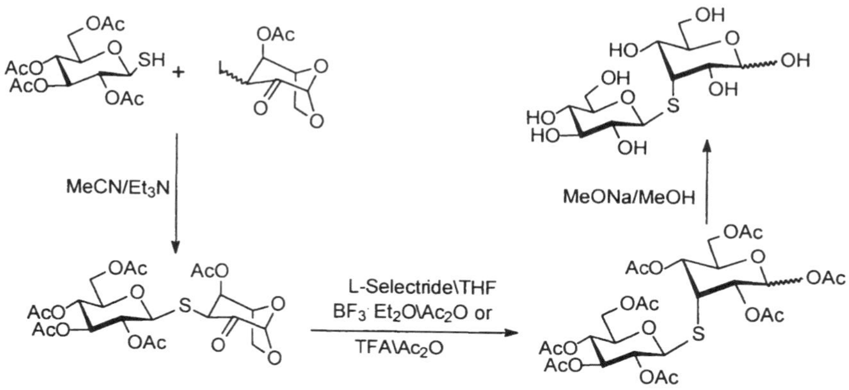

Scheme 4.

Additional modifications of saturated levoglucosenone derivatives can be achieved through additions to the carbonyl group at C-2. The addition of nitromethane and subsequent mesylation of the geminal secondary hydroxyl group followed by *in situ* elimination under basic conditions produces highly valuable nitroenones *(15)* (scheme 5). Both nitroalkenes exist as E/Z/ (1:1) isomeric mixture as detected by UV and NMR. Interestingly, attempts to separate the mixture by fractional crystallization using many polar solvents system failed and fast E/Z isomerization/eqilibration was always observed.

The conjugate system of the C-2 nitroalkenes should posses some interesting chemical reactivity and it should be an excellent Michael reaction acceptor with reactive nucleophiles. Moreover, the steric effect of the bulky 1,6-anhydro ring should be similar to that of levoglucosenone. As a consequence, nitroalkenes are excellent precursors for the stereoselective introduction of an additional sugar moiety at C-2 with subsequent additional functional group such as nitromethylene or its reduced/acetylated analog. Moreover, this unsaturated C-2 functionality additionally fixes the conformation of the system and most importantly sterically hinders the β-D-face of both enone molecules.

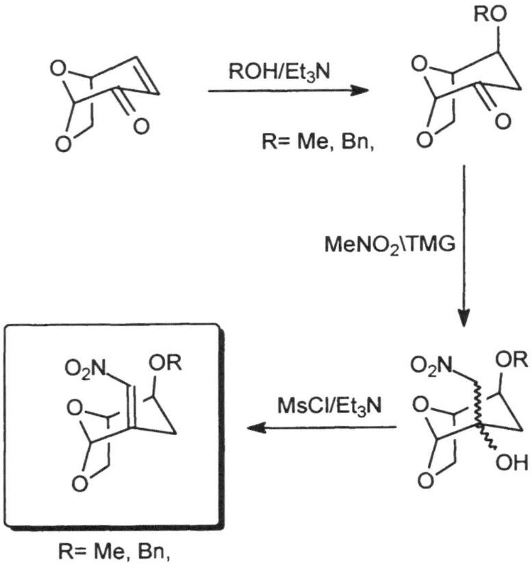

ROH/Et3N
R= Me, Bn,
RO
MeNO2\TMG
O2N
OR
OH
MsCl/Et3N
R= Me, Bn,

Scheme 5.

The reactivity of the nitroalkenes has been tested in the reaction with 1-thiosugars *via* conventional Michael reactions catalyzed by triethylamine. In both cases the stereoselective 1,2 addition proceeds by exclusive formation of an *exo*-adduct via formation of an S-linkage from the less hindered face of the molecule. As expected, the shielding effect of the 1,6-anhydro bridge effectively prevents the formation of the 2-equatorial product, yielding only the 2-axial products with a new quaternary center at C-2. This provides a stable molecule, as no epimerization or β-elimination is observed during the reduction of the nitro group.

exo-face attack
SugSH
OMe
NO2
SugSH
endo-face attack
RS
OMe
NO2

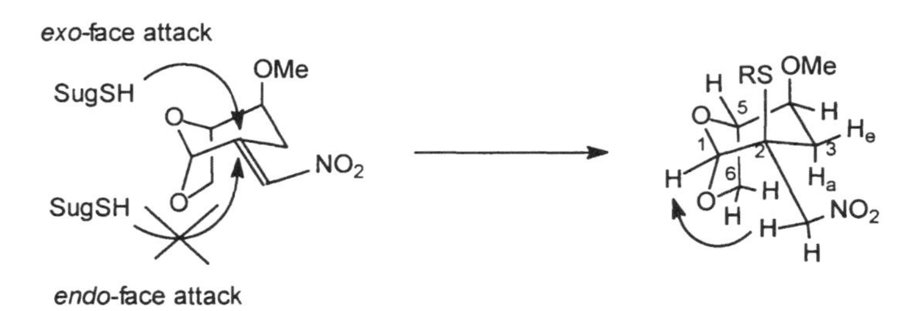

Figure 1. Stereochemistry of adduct and NOE effect between H at C-1 and H of nitromethyl at C-2

All the above factors clearly indicate the preferred stereochemistry of the adducts. The most direct way to prove the correct stereochemistry of the adducts is by measuring the coupling constants between H-3a and the $-CH_2-$ of the nitromethyl group at C-2, ie, $^3J_{CH}$= 2.8-3.2 Hz. The magnitude of these coupling constants strongly supports the proposed *gauche* arrangements with equatorial substituents at C-2. Additionally, a strong NOE effect is observed between the H at C-1 and one of the hydrogens on the nitromethylene group at C-2 further proves the correct stereochemistry at C-2. The ^{1}H NMR spectra of these adducts show a lack of coupling between H-4 and H-5, indicating that the pyranose ring of the adducts is in a $^1C_4$ conformation and is slightly distorted due to the presence of an equatorially oriented nitromethylene group at C-2 as illustrated in figure 1.

Consequently, the Michael addition reaction of sugar thiol proceeds smoothly with the formation of β- (1-2)-2,3-dideoxy-2-C-nitromethyl-thio-disacharides in 63-70 % yield (scheme 6).

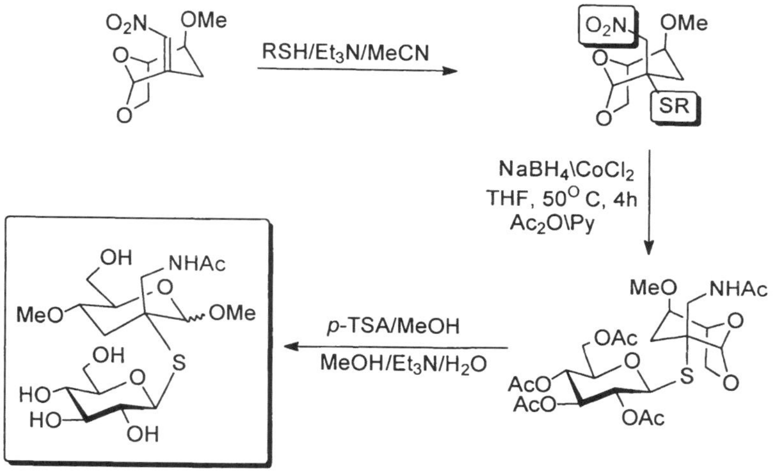

Scheme 6.

The reduction of the nitro group at C-2 of the thiodisacharides was efficiently carried out with sodium borohydride/cobalt chloride complex, followed by conventional acetylation. Final deprotection by ring opening was accomplished by the treatment with *p*-toluenesulfonic acid in methanol solution followed by deacetylation with aqueous/methanol solution containing catalytic amount of triethylamine.

This geminal type of functionality occurs when the sugar moiety is in specific stereo orientation, and with acetamido functionality. Additionally, the basic functional group (-NHAc) may act as a binding site with receptors. Such disaccharides should be valuable tools to probe any enzyme inhibitory activity of synthesized (1-2)-S-thiodisaccharides.

Again the stereochemistry of the new-branched thiodisaccharide was assigned on the basis of NOE results displaying a 5% enhancement between the C-acetamidomethyl group and the axial proton (3a-H) at C-3 and no enhancement of the 3e-H signal. The ^{13}C NMR signal of the methylene $-CH_2-$ group $\delta = 62.4$ at C-2 center is characteristic of the link with the quaternary C-2 and also clearly indicates the axial disposition of the new C-2 substituents.

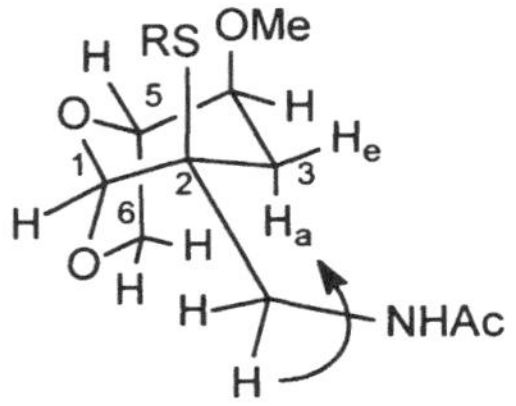

Figure 2. Stereochemistry and NOE correlation of 2-acetamido group effect between 3e-H at C-3 and H of nitromethyl at C-2

Reactivity of levoglucosenone as a dienophile in the Diels-Alder cycloaddition may be improved by introducing an electronegative group such as a halogen or nitro group. For that reason, the bromination of levoglucosenone has been studied in detail *(16)*. The predominant formation of 3-bromoglucosenone is always observed. Addition of bromine to levoglucosenone and concomitant elimination of hydrogen bromide with triethylamine facilitates a one-pot synthesis of 3-bromo-levoglucosenone (scheme 7).

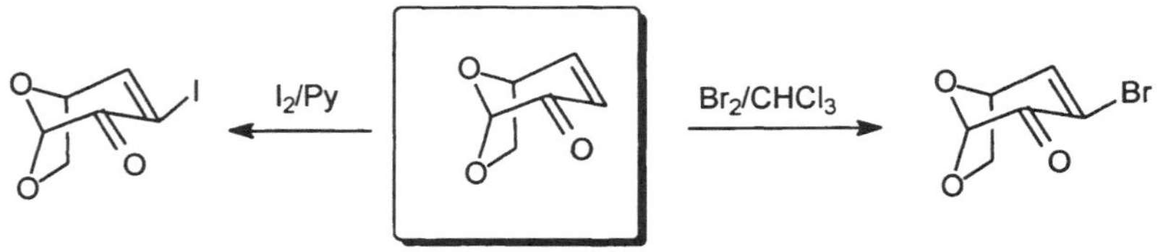

Scheme 7.

Addition of iodine to levoglucosenone has been conveniently performed by the treatment of this enone with a solution of iodine in anhydrous pyridine *(17)*, resulting in the formation of 3-iodolevoglucosenone in moderate (55%) yield.

The 3-nitro analog was also synthesized by Isobe laboratory with the intention of using it a chiral dienophile in synthetic approaches to heterocyclic systems of natural products, based on highly stereoselective cycloaddition reactions.

Isolevoglucosenone

Chemically named as 1,6-anhydro-2.3-dideoxy-β-D-glycero-hex-2-eno-pyranose-4-ulose, isolevoglucosenone is an alternative double dehydration product of levoglucosan (see scheme 1), however it was not detected among the products from acid-catalyzed pyrolysis of cellulose. This isomeric analog of levoglucosenone was first synthesized by Koll and coworkers *(18)* directly from levoglucosenone and from 1,6-anhydro-2, 3-O-isopropylidene-β-D-mannopyranose. Achmatowicz Jr and coworkers *(19)* synthesized racemic isolevoglucosenone from non-carbohydrate precursors. Furneaux and coworkers *(20)* synthesized isolevoglucosenone from levoglucosenone in six steps.

Our laboratory recently synthesized isolevoglucosenone directly from levoglucosenone *(21)* through four steps approach utilizing the key step of 2,3-sigmatropic rearrangement of an intermediate allylic selenide.

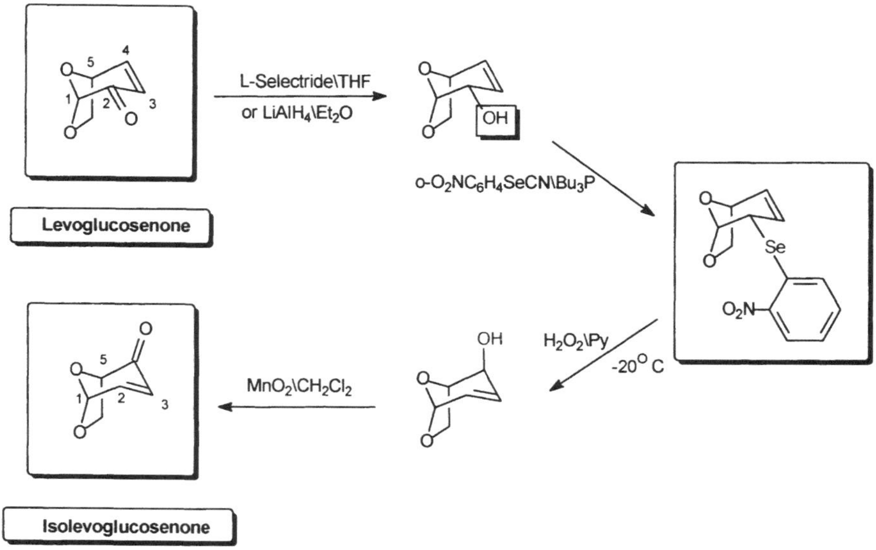

Scheme 8.

The [2,3]-sigmatropic shift leading to the rearrangement of the allylic selenide *via* the intermediate selenoxide during hydrogen peroxide oxidation is presumably catalysed by evolved o-nitrophenylseleninic acid. The mechanism of this sigmatropic rearrangement is shown in scheme 9. This key-step results in double bond transposition and introduction of allylic functionality at C-4 of isolevoglucosenone. To our knowledge, this is the first example of a [2,3]-sigmatropic rearrangement of a functionalized carbohydrate selenide.

Oxidation of the allylic alcohol was performed with manganese oxide in dichloromethane solution to produce isolevoglucosenone in high 89% yield.

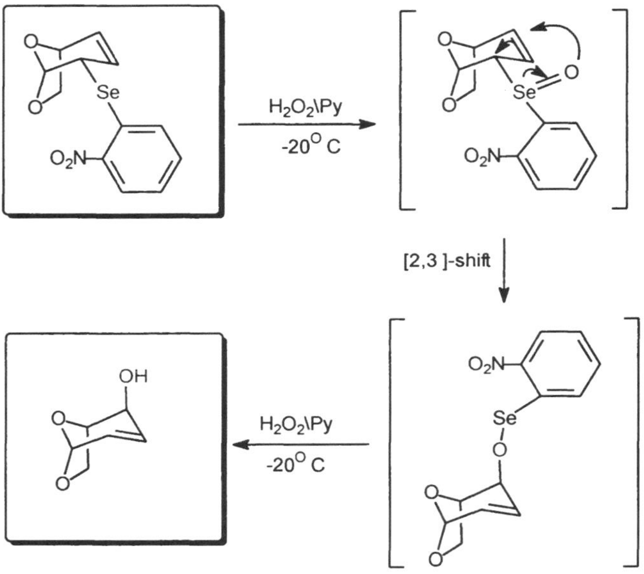

Scheme 9.

Among recent applications of isolevoglucosenone is the synthesis of new carbohydrate mimics, including C-disaccharides by the Baylis-Hillman type condensation of carbohydrate carbaldehydes with isolevoglucosenone as reported by Vogel and coworkers *(22-24)*. Horton and coworkers *(25)* also reported synthesis of isolevoglucosenone directly from 1,2:4,5-di-*O*-isopropylidene-3-*O*-methylsulfonyl-α-D-gluco-furanose and its application to the synthesis of biologically important deoxy aminosugars.

Our laboratory developed a new synthetic approach to (1-2)-*S*-thiodisaccharides *(26)* utilizing the reactivity of conjugated system of isolevoglucosenone. This synthetic approach (scheme 11) constitutes a general methodology, similar to our previously reported synthesis of (1-4)-*S*-3-deoxythidosacharides *(27-28)*.

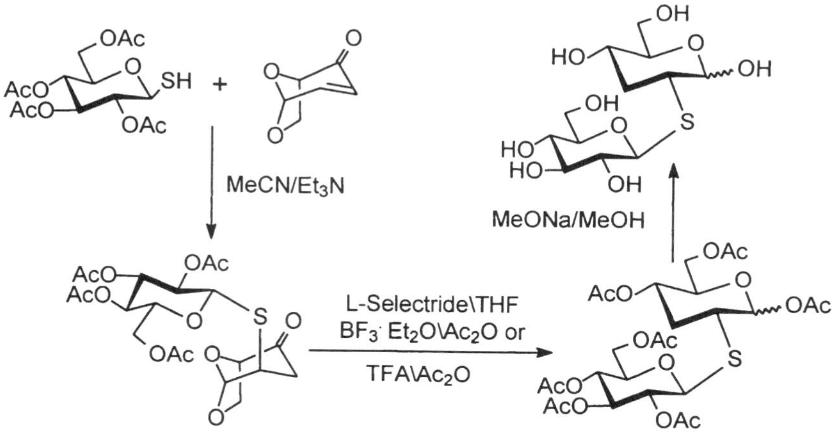

Scheme 10.

New Chiral Building Blocks from Isolevoglucosenone

Valuable analogs of functionalized isolevoglucosenone, particularly those similar to the levoglucosenone series bearing nitroalkenes functionality at C-4 deserve further consideration as new chiral precursors. They may be utilized in the synthesis of important classes of thioaminosugars having known biological activity. Indeed, these compounds are vital component of aminoglycoside antibiotics and for that particular reason fully deserve full synthetic exploration toward this new synthetic target.

Applying this new methodology of levoglucosenone functionalization at the C-2 position to isomeric isolevoglucosenone, we were able to successfully synthesize *(29)* new nitroalkenes with strategically important C-4 position for further functionalization at C-4 or C-3 positions. (Scheme 11)

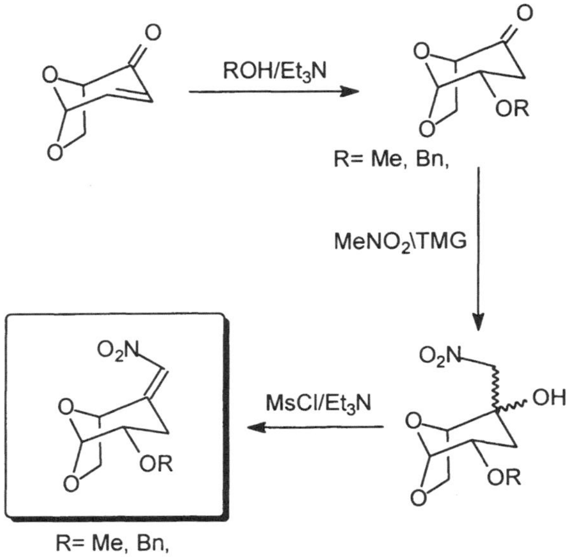

ROH/Et_3N

R= Me, Bn,

$MeNO_2$\TMG

$MsCl/Et_3N$

R= Me, Bn,

Scheme 11.

The high chemical reactivity of the conjugated system of levoglucosenone and the isomeric isolevoglucosenone is an excellent reason to explore new approaches for the synthesis of a variety of natural products targets that require stereoselective coupling with a sugar unit. As levoglucosenone and isolevoglucosenone are by far the most prominent carbohydrate molecules used in conjugate addition reactions, some of its tandem reactions involving the initial conjugate addition will be discussed in separate sections.

L-Arabinose Enones

Although developments in the chemistry of L-arabinose that use modern reagents as tools in organic synthesis produce only few universal functionalized building blocks, their potential value is enormous for further application as chiral organic material. An example is 3,4-O-isopropylidene acetal, which can be prepared simply and in very high yield from L-arabinose *(30)*.

New Chiral Building Blocks from L-arabinose

Natural L-arabinose as one of the highly functional pentoses with four chiral centers with different reactivities of secondary hydroxyls is an excellent precursor for the selective functionalization. Klemer and coworkers *(31)* synthesized one of the first valuable arabinose building block with protected C-1 and C-2 hydroxyl group and a conjugated enone between C-3, C-5 (scheme 12). This highly reactive enone should have synthetic potential through the introduction of additional functional groups at either C-3 or C-5.

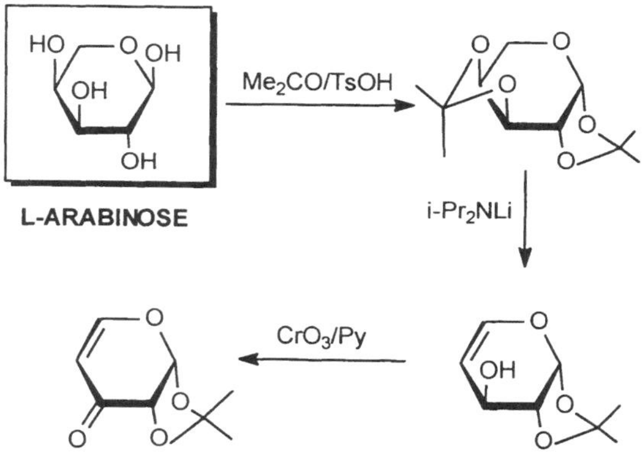

L-ARABINOSE

$Me_2CO/TsOH$

$i\text{-}Pr_2NLi$

CrO_3/Py

Scheme 12.

Indeed, stereospecific 1,4-additon of methyl lithium/copper iodide to the conjugated system of the above enone was initially reported by the authors*(31)*. We synthesized this convenient synthon and attempted to functionalize it further by removal of isopropylidene protecting group followed by acetylation at C-2. (scheme 13). All attempts failed due to extensive decomposition of the starting material, presumably through the β-elimination with formation of secondary polymerization products.

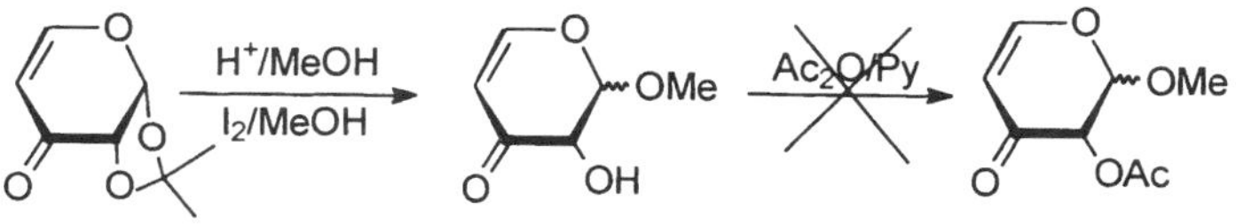

$H^+/MeOH$

$I_2/MeOH$

Ac_2O/Py

Scheme 13.

Our laboratory has also explored the synthetic utility of this chiral building block in the first synthesis of a new family of 3,5-diaminosugars*(32)*, as shown in scheme 14. The advantage of the above methodology is that a single step can be used for the simultaneous introduction of the amino functionality at both the C-3 and C-5 positions. Further examination of the chemistry of this universal enone is under development in our laboratory.

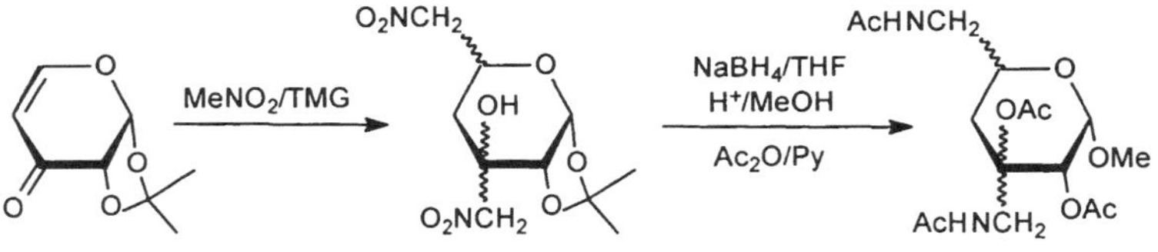

Scheme 14.

New perspectives

Recent developments in the chemistry of levoglucosenone during the last five years, as presented in this short review, will further the awareness of its potential in chemical syntheses and hopefully will encourage more extensive studies of this useful material in many different directions. The additional chiral functionality of levoglucosenone and its functionalized new synthons may create additional possibilities of research, not only in pure synthetic organic chemistry but also in polymer and combinatorial chemistry.

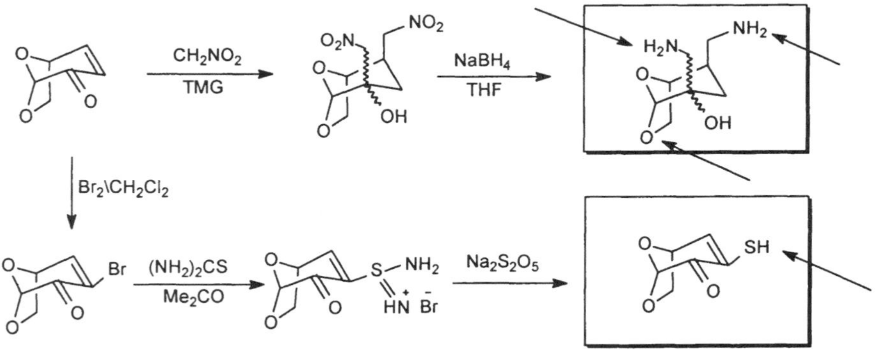

Scheme 15.

The most useful scaffolds would have modified functional groups such as– NH_2, -COOH, - SH, at C-2, C-3, C-4, and C-6. Our laboratory is developing a new family of levoglucosenone-based scaffolds with such functional groups at these positions. (Scheme 15).

The stereoselective, one-step synthesis of (1,2)-3-deoxy-thiodisaccharides *(26)* and (1,4)-3-deoxy-thiodisacharides *(27,28)* are classical examples of exploiting the excellent functionality of both levo- and isolevoglucosenone.

Many other laboratories *(2-11,33-45)* have made significant contributions to the chemistry of iso- and levoglucosenone. Further, interdisciplinary attempts to utilize the potential of both enones and their functionalized analogs in organic synthesis will be forthcoming.

Conclusion

Despite their availability and chiral richness, carbohydrates are still grossly underutilized as raw materials for fine chemistry. A number of new developments and synthetic methods have been devoted to this area of research during the last ten years, leading one to conclude that this is a rapidly growing field of carbohydrate chemistry. Despite the low level of pharmaceutical industry interest, chiral carbohydrate building blocks chemistry will likely be one of the frontiers in carbohydrate chemistry, especially in the area of small molecules and precursors for complex oligosaccharides of medicinal interest. The variety of methods for the functionalization of carbohydrate building blocks provides a number of stereoselective approaches to various classes of optically active derivatives, including sulfur and nitrogen heterocycles as well as rare carbohydrates.

Additionally, the environmental issue of utilizing waste cellulosic material and waste biomass products should be considered as an alternative green chemistry application to the production of many value added products. The combinatorial utilization of carbohydrate scaffolds based on chiral building block functionalization will also constitute attractive and relatively cheap starting materials. This rich selection of potential approaches, combined with further developments of new procedures and modern reagents, creates an enormous opportunity for the field to be at the frontier for many years to come.

References

1. For reviews see; *Levoglucosenone and Levoglucosans Chemistry and Applications* Witczak, Z. J. Ed. ATL Press Science Publishers; Mt. Prospect, IL **1994**; Witczak, Z. J. in *Studies in Natural Products Chemistry,* Atta-Ur-Rahman, Ed. Vol. 14, Elsevier Science Publishers,

Amsterdam, **1993**, pp. 267-282; Miftakhov, M. S.; Valeev, F. A.; Gaisina, I. N. *Uspekhi Khimi*, **1994**, *63*, 922; B. Becker, *J. Carb. Chem.* **2000**, *19*, 253; Witczak, Z. J. in *Chemicals and Materials from Renewable Resources,* Joseph J. Bozell, Ed. Vol. 784, ACS Symposium Series, Washington, DC, **2001**, pp. 81-97.

2. Tsuchiya, Y.; Swami, K. *J. Appl. Polym. Sci.* **1970**, *14,* 2003. Lam, L. K. M.; Fungi, D. P. C.; Tsuchiya, Y.; Swami, K. *J. Appl. Polym. Sci.* **1970**, *17,* 391.; Woodley, F. A. *J. Appl. Polym. Sci.* **1971**, *15*, 835.; Lipska, A. E.; McCasland, G. E. *J. Appl. Polym. Sci.* **1971**, *15*, 419.; Halpern, Y.; Riffer, R.; Broido, A. *J. Org. Chem.* **1973**, *38,* 204.; Broido, A.; Evett, M.; Hodges, C. C. *Carbohydr. Res.* **1975**, *44*, 267.; Ohnishi, A.; Takagi, E.; Kato, K. *Bull. Chem. Soc. Jpn.,* **1975**, *48*, 1956.; Koll, P.; Metzger, P. *Angew. Chem.* **1978**, *90*, 802.; Domburg, G.; Berzina, I.; Kupce, E.; Kirshbaum, I. Z. *Khim. Drev.* **1980**, 99. Domburg, G.; Berzina, I.; Kirshbaum, I. Z.; Gavars, M. *Khim.Drev.* **1978**, 105.; Halpern, Y.; Hoppech, J. P. *J. Org. Chem.* **1985**, *50,* 1556.; Smith, C. Z.; Chum, H. L.; Utley, J. H. P. *J. Chem. Res. Synopses* **1987**, *3*, 88.; Morin, C. *Tetrahedron Lett.* **1993**, 5095; Witczak, Z. J.; Mielguj, R. *Synlett* **1996**, 108: Witczak, Z. J.; Li, Y. *Tetrahedron Lett.* **1995**, 36, 2595; Witczak, Z. J.; Chhabra, R.; Chojnacki, J. *Tetrahedron Lett.* **1997**, *38*, 2215.

3. Shafizadeh, F.; Fu, Y. L. *Carbohydr.Res.* **1973**, *29*, 113.; Shafizadeh, F.; Chin, P. P. S. *Carbohydr.Res.* **1976**, *46,* 149.; Shafizadeh, F.; Chin, P. P. S. *Carbohydr.Res.* **1977**, *58,* 79.; Shafizadeh, F.; Furneaux, R. H.; Stevenson, T. T. *Carbohydr. Res.* **1979**, *71*, 169.; Ward D. D.; Shafizadeh F. *Carbohydr. Res.* **1981**, *93,* 284.; Ward, D. D.; Shafizadeh, F. *Carbohydr. Res.* **1981**, *95*, 155.; Shafizadeh, F.; Ward, D. D.; Pang, D. *Carbohydr. Res.* **1982**, *102*, 217.; Shafizadeh, F.; Furneaux, R. H.; Pang, D.; Stevenson, T. T. *Carbohydr. Res.* **1982**, *100*, 303.; Shafizadeh, F. *Pure Appl. Chem.,* **1983**, *55,* 705-720; Shafizadeh, F. *J. Anal. Appl. Pyrolysis,* **1982**, *3,* 283.; Shafizadeh, F.; Essig, M. G.; Ward, D. D. *Carbohydr.Res.* **1983**, *114,* 71.; Essig, M. G.; Shafizadeh, F. *Carbohydr. Res.* **1984**, *127*, 235.; Stevenson, T. T.; Furneaux, R. H.; Pang, D.; Shafizadeh, F.; Jensen, L. Stenkamp, R. E. *Carbohydr. Res.* **1983**, *112*, 179.; Stevenson, T. T.; Stenkamp, R. E.; Jensen, L. H.; Shafizadeh, F.; Furneaux. R. H. *Carbohydr. Res.* **1983**, *104*, 11.; Stevenson, T. T.; Essig, M. G.; Shafizadeh, F.; Jensen, L. H.; Stenkamp, R. E. *Carbohydr. Res.* **1983**, *118*, 261.; Essig, M. G.; Shafizadeh, F.; Cochran, T. G.; Stenkamp R. E. *Carbohydr. Res.* **1984**, *129*, 55.; Essig, M. G.; Stevenson, T. T.; Shafizadeh, F.; Stenkamp, R. E.; Jensen, L. H. *J. Org. Chem.* **1984**, *49*, 3652.;

4. Freskos, J. N.; Swenton J. *J. Chem. Soc. Chem. Commun.* **1984**, 658.; Swenton, J. S.; Freskos, J. N.; Dalidowicz, P.; Kerns. M. L. *J. Org. Chem.* **1996**, *61*, 459.
5. Taniguchi, T.; Nakamura, K.; Ogasawara, K. *Synlett* **1996**, 971.; Taniguchi, T.; Ohnishi, H.; Ogasawara, K. *Chem. Commun.* **1996**, 1477.; Takeuchi, M.; Taniguchi, T.; Ogasawara, K. *Synthesis* **1999**, 341.
6. Tolstikov, G.A.; Valeev, F.A.; Gareev, A. A.; Khalilov, L. M.; Miftakhov, M.S. *Zh. Org. Khim.* **1991**, *27*, 565.; Samet, A. V.; Laikhter, A. L.; Kislyi, V. P.; Ugrak, B. I.; Semenov, V. V. *Mendeleev Commun.* **1994**, 134.; Niyazymbetov, M. E.; Laikhter, A. L.; Semenov, V. V.; Evans, D. H. *Tetrahedron Lett.* **1994**, *35*, 3037.; Samet, A. V.; Kislyi, V. P.; Chernyshova, N. B.; Reznikov, D. N.; Ugrak, B. I.; Semenov, V. V. *Rus. Chem. Bull.* **1996**, *45*, 393.; Samet, A. V.; Yamskov, A. N.; Ugrak, B. I.; Vorontsova, L. G.; Kurella, M. G.; Semenov, V. V. *Rus. Chem. Bull.* **1996**, *45*, 393.; Valeev, F. A.; Gaisina, I. N.; Miftakhov, M. S. *Rus. Chem. Bull.* **1996**, 45, 2453.; Valeev, F. A.; Gaisina, I. N.; Sagitdinova, K. F.; Shitikova, O. V.; Miftakhov, M. S. *Zh. Org. Khim.* **1996**, *32*, 1365.; Samet, A. V.; Niyazymbetov, M. E.; Semenov, V. V.; Laikhter, A. L.; Evans, D. H. *J. Org. Chem.* **1996**, 61, 8786.; Miftakhov, M. S.; Gaisina, I. N.; Valeev, F. A. *Rus. Chem. Bull.* **1996**, *45*, 1942.; Laikhter, A. L.; Niyazymbetov, M. E.; Evans, D. H.; Samet, A.V.; Semenov, V. V. *Tetrahedron Lett.* **1993**, *34*, 4465.; Efremov, A. A.; Slaschinin, G. A.; Korniyets, E. D.; Sokolenko, V.A.; Kuznetsov, B. N. *Sibirskii Khim. Zhur.* **1992**, *6*, 34.; Efremov, A. A.; Konstantinov, A. P.; Kuznetsov, B. N. *J. Anal. Chem.* **1994**, *49*, 742.
7. Bahte, P.; Horton, D. *Carbohydr. Res.* **1983**, *122,* 189.; Bahte. P.; Horton, D. *Carbohydr. Res.* **1985**, *139,* 191.
8. Blake, A .J.; Forsyth, A. C.; Paton, M. R. *J. Chem. Soc. Chem. Commun.* **1988**, 440.; Dawson, I. M.; Johnson, T.; Paton, M. R.; Rennie, R.A.C. *J. Chem. Soc., Chem. Commun.* **1988**, 1339.; Blake, A. J.; Dawson, I. M.; Forsyth, A. C.; Gould, R.O.; Paton, M. R.; Taylor, D. *J. Chem. Soc. Perkin Trans. 1* **1993**, 75.; Blake, A. J.; Cook, T. A.; Forsyth, A. C.; Gould, R. O.; Paton, R. M. *Tetrahedron* **1992**, *48*, 8053.; Blake, A. J.; Gould, R. O.; Paton. R. M.; Taylor, P. G. *J. Chem. Res. Synopses* **1993**, 289.; Forsyth, A. C.; Paton, R. M.; Watt, A. *Tetrahedron Lett.* **1989**, 993.; Forsyth, A. C.; Gould, R. O.; Paton, R. M.; Sadler, I. H.; Watt, I. *J. Chem. Soc., Perkin Trans.1* **1993**, 2737.
9. Isobe, M.; Ichikawa, Y.; Goto, T. *Tetrahedron Lett.* **1981**, 4287.; Isobe, M.; Fukami, N.; Goto, T. *Chem. Lett.* **1985**, 71.; Isobe, M.; Nishikawa, T.; Pikul, S.; Goto, T. *Tetrahedron Lett.* **1987**, 6485.; Isobe, M.; Fukami, N.; Nishikawa,T.; Goto, T. *Heterocycles* **1987**, *25*, 521.; Isobe, M.; Fukuda, Y.; Nishikawa, T.; Chabert, P.; Kawai, T.; Goto, T. *Tetrahedron Lett.*

1990, 3327.; Isobe, M.; Nishikawa, T.; Yamamoto, N.; Tsukiyama,T.; Ino, A.; Okita, T. *J. Het. Chem.* **1992**, *29,* 619.; Nishikawa, T.; Araki, H.; Isobe, M. *Biosci. Biotechnol. Biochem.* **1998**, *62*, 190.;

10. Shibagaki, M.; Takahashi, K.; Kuno, H.; Honda, I.; Matsushita, H. *Chem. Lett.* **1990**, 307.; Ebata, T.; Matsumoto, K.; Yoshikoshi, H.; Koseki, K.; Kawakami, H.; Mashushita, H. *Heterocycles,* **1990**, *31*, 423.; Mashushita, H.; Hoshitake N.; Itoh, K. *Heterocycles* **1990**, *31*, 1585.; Matsumoto K.; Ebata T.; Koseki, K.; Kawakami, H.; Matsushita H. *Heterocycles* **1991**, *32*, 2225.; Matsumoto, K.; Ebata, T.; Koseki, K.; Kawakami, H.; Matsushita, H. *Bull. Chem. Soc. Jpn.* **1991**, *64,* 2309.; Matsumoto, K.; Ebata, T.; Koseki, K.; Okano, K.; Kawakami, H.; Matsushita, H. *Heterocycles* **1992**, *34*, 1935.; Koseki, K.; Ebata, T.; Kawakami, H.; Matsumoto, K.; Ebata, T.; Koseki, K.; Okano, K.; Kawakami, H.; Matsushita, H. *Carbohydr. Res.* **1993**, *246*, 345.; Matsumoto, K, Ebata, T.; Koseki, K.; Okano, K.; Kawakami, H. Matsushita, H. *Carbohydr. Res.* **1993**, *246*, 345.; Matsumoto, K.; Ebata, T.; Matsushita, H. *Carbohydr. Res.* **1995**, *279*, 93.
11. Gelas-Mialhe, Y.; Gelas, J.; Avenel, D.; Brahmi, R.; Gillie-Pandraud, H. *Heterocycles* **1986**, *24*, 931.
12. Gelas-Mialhe, Y.; Gelas, J. *Carbohydr.Res.* **1990**, *199,* 243.
13. Witczak, Z. J.; Kaplon, P.; Kolodziej, M. *Monatshefte Chem.* **2002**, *133*, 521.
14. Bekaert, A.; Barberan, O.; Gervais, M.; Brion J-D. *Tetrahedron Lett.* **2000**, *41*, 2903.
15. Witczak, Z. J.; Chhabra, R.; Boryczewski, D. *J. Carbohydr. Chem.* **2000**, *19*, 543.
16. Ward, D. D.; Shafizadeh, F. *Carbohydr. Res.* **1981**, *93*, 287.
17. Bamba, M.; Nishikawa, T.; Isobe, M. *Tetrahedron Lett.* **1996**, 37, 8199.
18. Koll, P.; Schultek, T.; Rennecke, R-W. *Chem Ber.* **1976,** *109*, 337.
19. Achmatowicz, Jr. O.; Bukowski, P.; Szechner, B.; Zwierzchowska, Z.; Zamojski, A. *Tertahedron,* **1971**, *27*, 1973.
20. Furneaux, R. H.; Gainsford, G. T.; Shafizadeh, F.; Stevenson, T. T. *Carbohydr. Res.* **1989**, *146,* 113.
21. Witczak, Z. J.; Kaplon, P.; Kolodziej, M. *J Carbohydr. Chem.* **2002**, *21*, 143.
22. Vogel, P. *Chimia,* **2001**, *55*, 359.
23. Zhu, Y-H.; Vogel, P. *Tetrahedron Lett.* **1998**, 39, 31.
24. Zhu, Y-H.; Demange, P.: Vogel, P. *Tetrahedron: Asymmetry,* **2000**, *11*, 263.; Zhu, Y-H.; Vogel, P. *J. Chem. Soc., Chem. Commun.* **1999**, 1873.; Zhu, Y-H.; Vogel, P. *J. Org.Chem.* **1999**, *64,* 666.
25. Horton, D.; Roski, J. P.; Norris, P. *J. Org. Chem.,* **1996**, *61,* 3783.
26. Witczak, Z. J.; Chen, H.; Kaplon, P. *Tetrahedron: Asymmetry,* **2000**, *11*, 519.

27. Witczak, Z. J.; Sun, J.; Mielguj, R. *Bioorg. Med. Chem. Lett.* **1995**, *5*, 2169.
28. Witczak, Z. J.; Chhabra, R.; Chen, H.; Xie, Q. *Carbohydr. Res.* **1997**, *301*, 167.
29. Witczak, Z. J.; Kolodziej, M. *Carbohydr. Res.* Submitted
30. Kiso, M.; Hasegawa, A. *Carbohydr. Res.* **1976**, *52*, 95.
31. Klemer, A.; Jung, G. *Chem Ber.* **1981**, *114*, 1192.
32. Witczak, Z. J.; Kolodziej, M. *Carbohydr. Res.* Submitted
33. Witczak, Z. J. 222rd Am Chem Soc Meeting, 2001, Abstract CARB 049
34. Furneaux, R. H.; Mason, J. M.; Miller, I. J. *J. Chem. Soc. Perkin Trans. 1* **1984**, 1923.
35. Koll, P.; Schultek, T.; Rennecke, R. W. *Chem. Ber.* **1976**, *109*, 337.
36. Griffin, A.; Newcombe, N. J.; Gallagher, T.; In *Levoglucosenone and Levoglucosans Chemistry and Applications* Witczak, Z. J. Ed. ATL Press Science Publishers; Mt Prospect, IL 1994; p.23.
37. Essig, M. G. *Carbohydr. Res.* **1986**, *156,* 225.
38. Becker, T.; Thimm, B.; Thiem, J. *J. Carbohydr. Chem.* **1996**, *15*, 1179.
39. Chew, S.; Ferrier, R. J. *J. Chem. Soc. Chem. Commun.* **1984**, 911.
40. Chew, S.; Ferrier, R. J.; Sinnwell, V. *Carbohydr. Res.* **1988**, *174*, 161.
41. Ferrier, R.; Furneaux, R. *Aust. J. Chem.* **1980**, *33,* 1025.
42. Ferrier, R.; Tyler, P. C. *J. Chem. Soc. Perkin Trans.1* **1980**, 2767.
43. Mori, M.; Chuman, T.; Kato, K.; Mori, K. *Tetrahedron Lett.* **1982**, 4593.
44. Mori, M.; Chuman, T.; Kato, K. *Carbohydr. Res.* **1984**, *129*, 73.
45. Blattner, R.; Page, D. M. *J. Carbohydr. Chem.* **1994**, *13*, 27.
46. Gomez, M.; Quincoces, J.; Peske, K.: Michalik, M. *J. Carbohydr. Chem.* **1999**, *18*, 851.

Chapter 2

A Convenient Procedure for the Preparation of Levoglucosenone and Its Conversion to Novel Chiral Derivatives

Walter S. Trahanovsky, Jason M. Ochaoda, Chen Wang, Kevin D. Revell, Kirk B. Arvidson, Yu Wang, Huiyan Zhao, Steven Chung, and Synthia Chang

Department of Chemistry, Iowa State University, Ames, IO 50011–3111

As part of a study to develop methods to obtain high-value nonracemic chiral compounds from biomass, we have developed a convenient method for converting cellulose to levoglucosenone in >10% yield. This procedure and methods to convert levoglucosenone into potentially useful chiral derivatives are presented.

Practical methods for deriving economically useful fuels and chemicals from renewable plant material, biomass, are desirable since they make it possible to use biomass instead of petroleum as the source of fuels and chemicals. There has been considerable effort focused on developing biological, chemical, and pyrolytic methods to convert biomass to useful fuels and chemicals.[1] A major component of biomass is cellulose (**1**) and the major products from the pyrolysis of cellulose (**1**) are charcoal, water, and small organic molecules, some of which are chiral but most of which are highly unsaturated and achiral.[1d,2] We proposed that the pyrolysis of cellulose (**1**) and other biomass materials in the presence of hydrogen-rich materials might allow reactive species produced at high temperatures to capture hydrogen atoms from the hydrogen-rich material to form small molecules that are more saturated.

Since biomass is chiral, these saturated molecules could also be chiral and could be useful chiral synthetic building blocks, so-called "chirons".[3] In addition to possibly producing useful chemicals, the transfer of hydrogen to the biomass material could also lead to a higher quality, more hydrogen-rich liquid fuel.

In an effort to observe this proposed transfer of hydrogen, a mixture of cellulose (**1**), soybean oil, and an acid catalyst was pyrolyzed at 300°C using the procedure described by Morin.[4] Analysis of the products showed no new significant products derived from the biomass but the yield of levoglucosenone (**2**), the major small organic compound produced under these conditions, was increased significantly.

cellulose (1) **levoglucosenone (2)**

On the basis of this observation, the pyrolysis of mixtures of cellulose (**1**) and various vegetable oils under a variety of conditions was studied, and we have developed a convenient method for converting cellulose (**1**) to levoglucosenone (**2**) in >10% yield.[6] The previously reported[7] yields of levoglucosenone (**2**) produced by the pyrolysis of cellulose (**1**) under acidic conditions are usually 2-5% and utilize special vessels and procedures that are more involved than ours. Our procedure is essentially a vacuum distillation.

Our method produces levoglucosenone (**2**) in a relatively pure form. In our procedure, cellulose (**1**) and an acid catalyst are added to a vegetable oil, such as soybean oil (in a 1:3 ratio of cellulose:oil), and the mixture is rapidly heated to 300 °C under vacuum. Within seconds levoglucosenone (**2**), water, and charcoal begin to form and the water and levoglucosenone (**2**) distill from the mixture and are condensed. The levoglucosenone (**2**) obtained from this procedure is relatively pure (*ca.* 75%) and can be further purified by distillation to give levoglucosenone (**2**) that is >80% pure. We can conveniently obtain multigram quantities of levoglucosenone (**2**) by this procedure and feel confident that the procedure could be scaled up to produce even larger quantities.

The following is the specific procedure for the preparation of levoglucosenone (**2**) by the pyrolysis of 5 g of cellulose (**1**).[8] To a 50-mL round-bottom flask fitted with a vacuum distillation apparatus was added phosphoric acid (25 mg, 0.5 wt %), cellulose (5 g) and vegetable oil (15 g). The slurry was stirred for about 5 min under reduced pressure (20-30 mm Hg), and then heated by an appropriate heating mantle for 7 min to 270°C, as indicated by the internal thermometer. The reaction mixture began to turn black and water distillate appeared at about 140°C. Yellow distillate containing water and levoglucosenone (**2**) appeared on the flask wall around 270°C, and the temperature in the distillation head reached around 110-120°C. The reaction temperature was increased to 300-310°C over 15 min until no more distillate came over. The yellow distillate was extracted by methylene chloride, and the methylene chloride solution was dried over $MgSO_4$. After rotary evaporation of the solvent, levoglucosenone (**1**) can be used directly, or further purified by distillation. This procedure produces several hundred milligrams of levoglucosenone (**2**).

We have also studied the use of paper as the source of cellulose (**1**) and we have learned that if the paper is pretreated with acid, yields of *ca.* 5% levoglucosenone (**2**) can be obtained. The paper was preacidified following the procedure of Shafizadeh and Chin.[9] A quantity of 25 g of paper from a newspaper was shredded into strips which were then cut by hand into smaller pieces. To a 1000-mL single-neck round-bottom flask were added the shredded paper, 180 mL of water, and 0.8 g (3.2% of the weight of the paper) of 88% phosphoric acid. The mixture was heated for 3 hours at 60-70°C and then the water was removed by using a rotary evaporator (4 hr). The preacidified paper was then mixed with 25 g of soy oil and the mixture was pyrolyzed in a 500-mL round-bottom flask; the yield as determined by GC using octyl alcohol as the internal standard was ca. 5%.

Conversion of levoglucosenone (2) to chiral derivatives.

The potential of levoglucosenone (**2**) for use in organic synthesis is exceedingly high.[9] It is a relatively small (six carbon atoms), nonracemic chiral, rigid molecule with several important functional groups including a ketone group, a double bond conjugated with the ketone, a protected aldehyde, and two protected hydroxyl groups.[5] We have converted levoglucosenone (**2**) to several derivatives which have potential for use in the synthesis of complex nonracemic chiral compounds. Each of these derivatives has been fully characterized by infrared, mass, and ^1H and ^{13}C NMR spectroscopy.[8,10]

Cycloaddition reactions of levoglucosenone (**2**) have been widely investigated and include the Diels-Alder reaction with various dienes such as cyclopentadiene, butadiene, and isoprene.[11-14] In addition, 1,3-dipolar cycloadditions have been investigated.[15]

We studied the Diels-Alder reaction of levoglucosenone (**2**) with the very reactive diene 2,3-dimethylene-2,3-dihydrofuran (**3**),[10] the furan-based *ortho*-quinodimethane, which is readily available by the Flash Vacuum Pyrolysis (FVP) of ester **4**.[16] Diene **3** dimerizes very rapidly[16]

CH_2–O–C(=O)–Ph
CH_3
4
FVP
-PhCOOH
CH_2
CH_2
3

and the thermal reaction of **2** and **3** gave only 12% yield of a mixture of two isomeric cycloadducts, **5** and **6**. Addition of the Lewis acid boron trifluoride[17] increased the rate of the Diels-Alder reaction and the yield of products **5** and **6** increased to 40%.[10]

3 + 2 —BF_3•OEt_2→ 5 + 6

The thermal reaction produced **5** and **6** in a 3:2 ratio but the Lewis acid-catalyzed reaction gave a 1:1 ratio. We could not separate the two cycloadducts, but based the indicated constitution on analysis of the mass spectrum and ^{1}H NMR spectrum of the mixture of the two compounds. The stereochemistry was assigned on the basis that the Diels-Alder reaction is normally suprafacial-suprafacial[18] and that for other cycloaddition reactions of levoglucosenone (**2**) the addition anti to the 1,6-anhydro bridge is favored.[13-15]

Michael additions[19] to levoglucosenone (**2**) have been reported to go with high stereoselectivity. Only products derived by addition anti to the 1,6-anhydro bridge, exo products, have been reported for the Michael addition of methyl and *n*-pentyl cuprates[20] and other Michael donors[21] to levoglucosenone (**2**).

We have found that the Michael addition of *n*-hexyl cuprate to levoglucosenone (**2**) gives the exo adduct **7** (stereochemistry determined by NOE ^{1}H NMR spectroscopy) as the major product in 70-80% yield; less than

2 + $[CH_3(CH_2)_5]_2CuLi$ —THF, 2-4h, -78 °C → 25 °C→ **7**

C_6H_{13}, H

7 (major isomer, yield 70-80%)

5% of the endo isomer is formed. We have studied compound **7** as a source of chiral derivatives because a) compound 7 is a simple ketone without a conjugated double bond which makes it less reactive than levoglucosenone (**2**) and b) the large aliphatic group, the hexyl group, leads to organic soluble compounds that are easier to isolate. Our methods allow us to prepare multigram quantities of very pure (>98%) hexyl adduct 7.

Baeyer-Villiger oxidation of ketone 7, an oxidation with *meta*-chloroperbenzoic acid, gave a very good yield of a mixture of formates **8** and **9**

8 **9**

8/9 = 2.4/1; yield 70-90%

in a ratio of 2.4:1.0 (the sample of **7** may have contained a small amount, <5%, of the endo isomer but no products from the endo isomer were detected). The mixture of **8** and **9** was converted to γ-lactone **10** by treatment with LiOH.

8 + **9** —LiOH, THF, reflux→

10
85%

The stereochemistry of compounds **8**, **9**, and **10** was determined by NOE spectroscopy. It has been reported [20a] that the methyl and pentyl adducts of levoglucosenone (**2**) when oxidized by peracetic acid give γ-lactones that correspond to lactone **10**, but no intermediate formates were reported.

Baeyer-Villiger oxidation of a cyclic ketone normally gives a ring-expanded lactone[22] so lactone **11** is the expected Baeyer-Villiger product of **7**, not formates **8** and **9**. This unexpected observation may be due to the fact that 7 has a ketal group α to the carbonyl group. Levoglucosenone (**2**) also has this structural feature and the Baeyer-Villiger oxidation of it has been reported to yield a formate as a major product.[20a] It is conceivable that **11** is formed initially but undergoes acid-catalyzed rearrangements to **8** and **9**. Alternatively, the initially formed peracid adduct of **7**, **12**, may go directly to the formates.

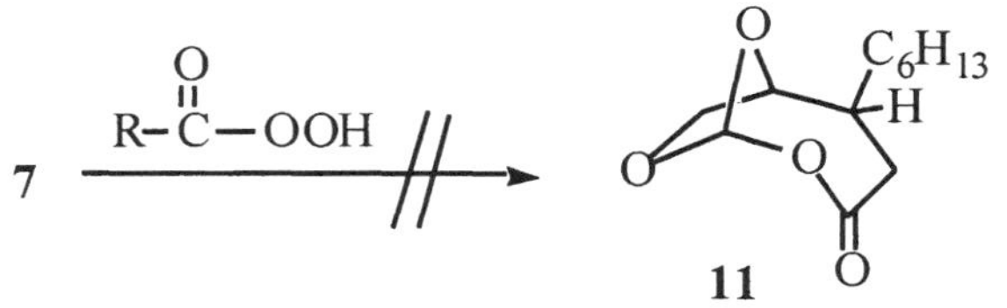

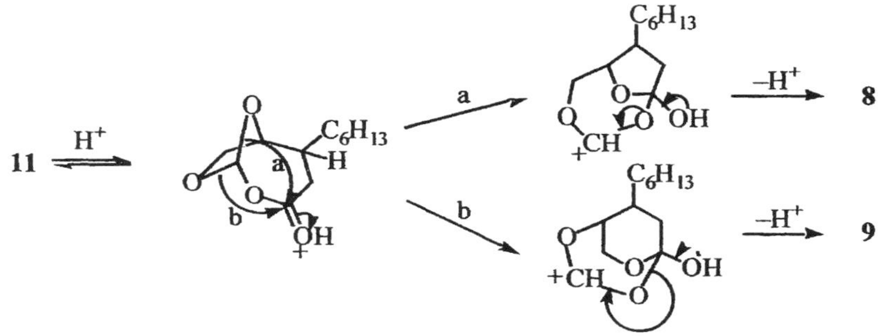

Possibly each formate is derived from a specific stereoisomer of **12**, **12-exo**, and **12-endo**.

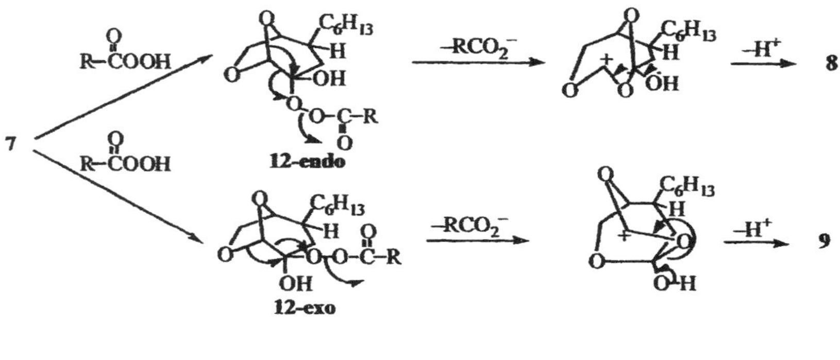

The Baeyer-Villiger oxidation of ketone 7 by the procedure of Grieco[23] which uses H_2O_2 in acetic acid also gave formates **8** and **9** in a ratio similar to that obtained when *m*-chloroperbenzoic acid was used. Also, it was noted that limited treatment of the formate mixture with LiOH resulted in formation of a mixture of lactone **10** and another compound. Spectroscopic evidence suggests that this second product is the δ-lactone formed by removal of the formate group from **9**. Evidently, prolonged treatment with LiOH results in isomerization of this δ-lactone to γ-lactone **10**.

The carbonyl group of 7 is readily reduced by sodium borohydride[24] to give a mixture of the epimeric alcohols **13** and **14**. This mixture of epimeric

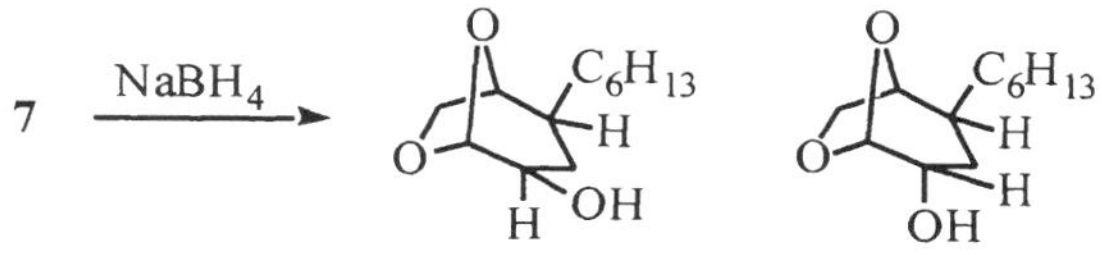

7 $\xrightarrow{NaBH_4}$ **13** + **14**

ratio 55/45; yield 100%

alcohols can be converted to olefin **15** via the mixture of mesylates **16**.[25] Compounds **13** to **16** should be useful synthetic building blocks since with their

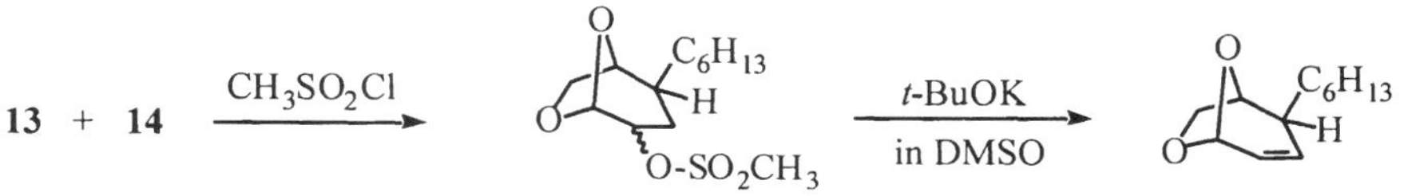

13 + **14** $\xrightarrow{CH_3SO_2Cl}$ **16** $\xrightarrow[\text{in DMSO}]{t\text{-BuOK}}$ **15**

simple, common functional groups, they can be converted to many other enantiomerically pure compounds.

The hexyl derivative 7 was pyrolyzed[26] in the gas phase (Flash Vacuum Pyrolysis conditions)[27] with the hope of losing CO to produce **17**.[28] Little or no **17** was produced, but a reasonable yield of several products including two isomers of 7, always in a ratio of 4:1, was obtained. We have determined (using

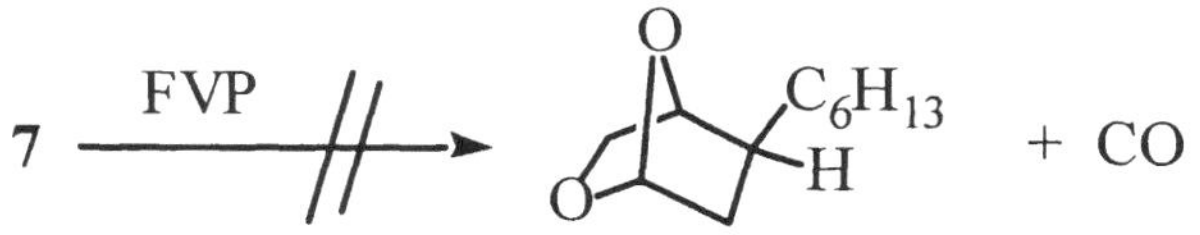

7 —FVP—//→ **17** + CO

infrared, mass, and ^{1}H, ^{1}H NOE, and ^{13}C NMR spectroscopy) that these compounds are **18** and **19**. We propose that these two bicyclic lactones are formed through a 6-electron rearrangement in which the carbonyl oxygen atom closes to form an ether linkage, the acetal group opens to form an ester, and a hydrogen atom shifts from the acetal carbon atom to the neighboring carbonyl

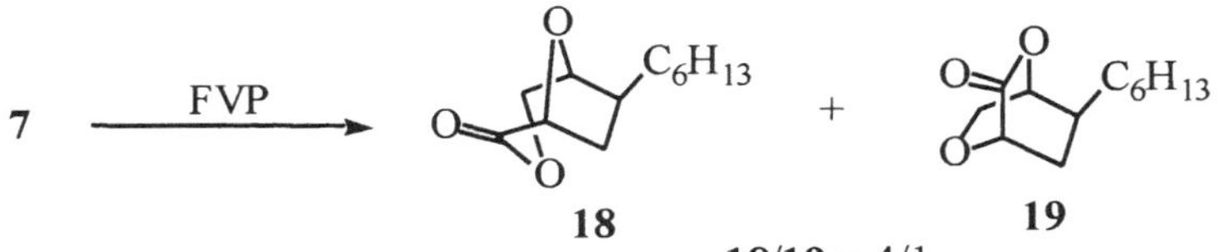

18/19 = 4/1

carbon atom. The 6-membered ring of the transition state for the formation of **18** is in a chair conformation but in the transition state for the formation of **19**, the 6-membered ring is in the boat conformation. This difference may account

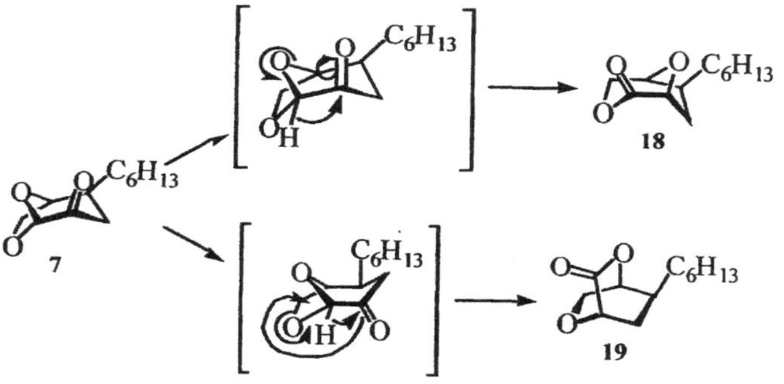

for the higher yield of **18** relative to **19**. Although electrocyclizations involving heteroatoms have been extensively studied,[29] we are not aware of any precedent for this type of rearrangement.

Compound **7** was pyrolyzed under a variety of conditions with the objective of optimizing the yields of **18**, **19**, and other products. In addition to **18** and **19**, it was found that another product is produced, especially at high temperatures. Spectroscopic data indicate that this product is aldehyde **20**. The mechanism for the formation of **20** is not clear at this time.

$$CH_2{=}C(C_6H_{13})\text{-}C(H){=}O$$

20

Acknowledgements

We thank the Iowa Energy Center for generous financial support and Qiqing (Max) Zhong for help with literature references.

References

1. For example: (a) Bungay, H. R. *Energy, The Biomass Options*, John Wiley & Sons: New York, **1981**, 113-129. (b) *Biomass Conversion Technology*, M. Moo-Young, Ed., Pergamon Press: New York, **1987**. (c)*Pyrolysis Oils from Biomass*, E. J. Soltes and T. A. Milne, Eds., American Chemical Society: Washington, DC, **1988**. (d) *Biomass Pyrolysis Liquids: Upgrading and Utilization*, A. V. Bridgwater and G. Grassi, Eds., Elsevier Applied Sciences: New York, **1991**. (e) *Bioconversion of Forest and Agricultural Plant Residues*, J. N. Saddler, Ed., C•A•B International: Wallingford, UK, **1993**. (f) *Developments in Thermochemical Biomass Conversion*, Volumes 1 and 2, A. V. Bridgwater and D. G. B. Boocock, Eds., Blackie Academic & Professional, New York, **1997**. (g) Davidson, B. H.; Finkelstein, M.; Wyman, C. E. *Biotechnology for Fuels and Chemicals*, Humann Press: New Jersey, **1997**.
2. (a) Halpern, Y.; Riffer, R.; Brodio, A. *J. Org. Chem.* **1973**, *2*, 204. (b) Faix, O.; Fortmann, I.; Bremer, J.; Meier, D. *Holz als Roh- und Werkstoff.* **1991**, *49*, 299.
3. (a) *Total Synthesis of Natural Products: The Chiron Approach*, S. Hanessian, Ed., Pergamon Press: Oxford, New York, **1983**. (b) *Preparative Carbohydrate Chemistry*, S. Hanessian, Ed., Marcel Dekker: New York, **1994**.
4. Morin, C. "A Simple Bench-Top Preparation of Levoglucosenone" in ref. 5, 17.
5. *Levoglucosenone and Levoglucosans*, Z. J. Witczak, Ed., ATL Press, Inc. Science Publishers: Mt. Prospect, Illinois, **1994**.

6. Preliminary report of this procedure: "Convenient procedure for the preparation of levoglucosenone from cellulose and the conversion of levoglucosenone to novel chiral derivatives", paper by W. S. Trahanovsky, C. Wang, J. M. Ochoada, and S. Chang as part of the Symposium on Chemistry of Renewable Fuels and Chemicals, at the 217th National Meeting of the American Chemical Society, Anaheim, California, March 1999. Abstract and Preprint FUEL 35.
7. The preparation of levoglucosenone (**2**) is mentioned in several chapters in ref. 5
8. Wang, Chen, Ph.D. Dissertation, Iowa State University, Ames, Iowa, **1999**.
9. Shafizadeh, F.; Chin, P. P. S. *Carbohydr. Res.* **1977**, *58*, 79-87.
10. Ochoada, J. M., M.S. Thesis, Iowa State University, Ames, Iowa, **1999**.
11. (a) Valeev, F. A.; Gaisina, I. N.; Miftakhov, M. S. *Ivz. Akad. Nauk. Ser. Khim.* **1996**, *10*, 2584-2585. (b) Miftakhov, M. S.; Valeev, F. A.; Gaisina, I. N.; Shitikova, O. V. *Zh. Org. Khim.* **1993**, *29*, 1122-37. (c) Tolstikov, G. A.; Miftakhov, M. S.; Valeev, F. A., Gareev, A. A. *Zh. Org. Khim.* **1990**, *26*, 2461-2. (d) Isobe, M.; Nishikawa, T.; Pikul, S.; Goto, T. *Tetrahedron Lett.* **1987**, *28*, 6485-8. (e) Isobe, M.; Fukami, N.; Nishikawa, T.; Goto, T. *Heterocycles* **1987**, *25*, 521-32. (f) Isobe, M.; Fukumi, N.; Naoki, G.; Goto, T. *Chem. Lett.* **1985**, *1*, 71-4. (g) Ferrier, R. J. *J. Chem. Soc., Chem. Commun.* **1984**, *14*, 911-12. (h) Bhate, P.; Gallucci, J.; Horton, D. *Cryst. Struct. Commun.* **1984**, *3*, 468-70. (i) Shafizadeh, F.; Essig, M. G.; Ward, D. *Carbohydr. Res.* **1983**, *114*, 71-82. (j) Ward, D. D.; Shafizadeh, F. *Carbohydr. Res.* **1981**, *95*, 155-76.
12. Bhate, P.; Horton, D. *Carbohydr. Res.* **1983**, *122,* 189-99.
13. (a) Isobe, M.; Yamamoto, N.; Nishikawa, T. *Front. Biomed. Biotechnol.* **1992**, *2*, 99-108. (b) Isobe, M.; Yamamoto, N.; Nishikawa, T., ref. 5, 99.
14. Miftakhov, M. S.; Gaisina, I. N.; Valeev, F. A. *Ivz. Akad. Nauk. Ser. Khim.* **1996**, *8*, 2047-2049.
15. (a) Wiest, H.; Long, D.; Sauer, V. J. *Z. Naturforsch.* **1962**, *17b*, 206. (b) Blake, A. J.; Forsyth, A. C.; Hewitt, B. D. *J. Chem. Soc., Perkin Trans I* **1987**, 2371.
16. (a) Trahanovsky, W. S.; Cassady, T. J.; Woods, T. L. *J. Am. Chem. Soc.* **1981**, *103*, 6691. (b) Chou, C.-H.; Trahanovsky, W. S. *J. Org. Chem.* **1986**, *51*, 4208. (c) Chou, C.-H.; Trahanovsky, W. S. *J. Am. Chem. Soc.* **1986**, *108*, 4138. (d) Chou, C.-H.; Trahanovsky, W. S. *J. Org. Chem.* **1995**, *60*, 5449. (e) Trahanovsky, W. S.; Cassady, T. J.; Chou, C.-H. *J. Org. Chem.* **1994**, *59*, 2613.
17. Wenkert, E.; Friguelli, F.; Taticchi, A. *Org. Prep. And Proceed.* **1990**, *22(2)*, 131.

18. Oppolzer, W. *Comprehensive Organic Synthesis*, Trost, B. M., Ed.; Pergamon Press, Chapter 4, **1991**.
19. (a) Rossister, B. E.; Swingle, N. M. *Chem. Rev.* **1992**, *92,* 771. (b) Lipshutz, B. H.; Senguptas, S. *Org. React.* **1992**, *41*, 135. (c) Weinmann, H.; Winterfelt, E. *Synthesis* **1995**, 1097. (d) Lipshutz, B. H.; Aue, D. H.; James, B. T. *Tetrahedron Lett.* **1996**, *37*, 8371. (e) Yamamoto, K.; Ogura, H.; Jukuta, J.; Inone, H.; Hamada, K.; Sugiyama, Y.; Yamada, S. *J. Org. Chem.* **1998**, *63*, 4449.
20. (a) Ebata, T.; Koseki, K.; Okano, K.; Matsushita, H., ref. 5, 59. (b) Ebata, T.; Matsumoto, K.; Yoshikoshi, H.; Koseki, K.; Kawakami, H.; Okano, K.; Matsushita, H. *Heterocycles* **1993**, *36*, 1017.
21. (a) Shafizadeh, F.; Ward, D. D.; Pang, C. *Carbohydr. Res.* **1982**, *102*, 217.
22. (a) Renz, M.; Meunier, B. *Fr. Eur. J. Org. Chem.* **1999**, *4*, 737. (b) Allen, J. L.; Paquette, K. P.; Porter, N. A. *J. Am. Chem. Soc.* **1998**, *120*, 9362. (c) Gottlich, R.; Yamakoshi, K.; Sasai, H.; Shibasaki, M. *Synlett* **1997**, 971. (d) Krow, G. R. *Tetrahedron* **1981**, *37*, 2697. (e) Sosnowski, J. J.; Danaher, E. B.; Murry, R. K., Jr. *J. Org. Chem.* **1985**, *50*, 2759. (f) Plesnicar, B. *Oxidation in Organic Chemistry*, Trahanovsky, W. S., Ed., Academic Press: New York, **1978**, 254.
23. Grieco, P. A. *J. Org. Chem.* **1972**, *37*, 2363.
24. Blattner, R.; Page, D. M. *J. Carbohydr. Chem.* **1994**, *13*, 27.
25. Trahanovsky, W. S.; Fischer, D. R. *J. Am. Chem. Soc.* **1990**, *112*, 4971.
26. Revell, K. D., M.S. Thesis, Iowa State University, Ames, Iowa, **1999**.
27. Trahanovsky, W. S.; Ong, C. C.; Pataky, J. G.; Weitl, F. J.; Mullen, P. W.; Clardy, J. C.; Hansen, R. S. *J. Org. Chem.* **1971**, *36*, 3575.
28. For reviews, see (a) McNab, H. *Contemporary Org. Synth.* **1997**, 373. (b) Brown, R. F. C. *Pure & Appl. Chem.* **1990**, *62*, 1981. (c) Cadogan, J. I. G.; Hickson, C. L.; McNab, H. *Tetrahedron* **1986**, *42*, 2135. (d) Karpf, M. *Angew. Chem. Int. Ed. Engl.* **1986**, *25*, 414. (e) Wiersum, U. E. *Recl. Trav. Chim. Pay.-Bas.* **1982**, *101*, 317, 365. (f) Brown, R. F. C. *Pyrolysis Methods in Organic Chemistry*, Academic: New York, **1980**, Chapter 2.
29. For a review on the synthesis of heterocyclic systems through concerted mechanism, see Carruthers, W. *Some Modern Methods of Organic Synthesis,* Cambridge: Cambridge University Press, **1993**, Chapter 4.

Chapter 3

Preparation and Exploitation of an Artificial Levoglucosenone

Kunio Ogasawara

Pharmaceutical Institute, Tohoku University, Aobayama, Sendai 980–8578, Japan

Levoglucosenone and its functionalized analogue have been synthesized from furan by employing either enzymatic resolution or asymmetric synthesis. The potential of the latter as a chiral building block has been demonstrated by synthesis of all eight hexose diastereomers and some other natural products.

Introduction

Levoglucosenone[1] (–)-**1** is a pyrolysis product of cellulose having a 7,8-bicyclo[3.2.1]oct-3-en-2-one framework. Because of its high functionality confined in the biased framework exerting inherent convex-face selectivity, it has received considerable interest as a versatile chiral building block for diastereocontrolled construction of natural products.[2] It is, however, still not fully used owing to its less effective production limited to the particular enantiomer originated from cellulose. Moreover, its versatile utility is precluded by the presence of a rather sturdy internal acetal functionality, the cleavage of which required rather strong conditions. Therefore, development of an efficient production of both enantiomers of levoglucosenone **1** as well as its synthetic equivalent allowing facile acetal cleavage would greatly promote chiral synthesis. I will present here recent advances achieved in our laboratory involving the synthesis of levoglucosenone **1** itself and its functionalized analogues in both enantiomeric forms without using sugar precursors, and the exploitation of the latter in the synthesis of aldohexoses and other natural products.

Synthesis of Levoglucosenone

Besides the cellulose pyrolysis producing levoglucosenone (–)-**1**, syntheses of both enantiomers of **1** have been reported.[1,3] However, these methods use natural carbohydrate precursors and the results are practically less than satisfactory. We, therefore, explored first the synthesis of levoglucosenone **1** without using a naturally occurring starting material.

In order to obtain levoglucosenone **1** in both enantiomeric forms by employing lipase-mediated kinetic resolution, we used acrolein dimer **2** as the starting material.[4] **2** was first transformed to the bicyclic ketone (±)-**6** by sequential four steps of reactions *via* **3~5**. Racemic levoglucosenone (±)-**1** was obtained from **6** *via* the silyl ether **7** by employing the Saegusa reaction. To carry out lipase-mediated resolution, (±)-**1** was transformed into the *endo*-alcohol (±)-**8** and the acetate (±)-**9** (**Scheme 1**).

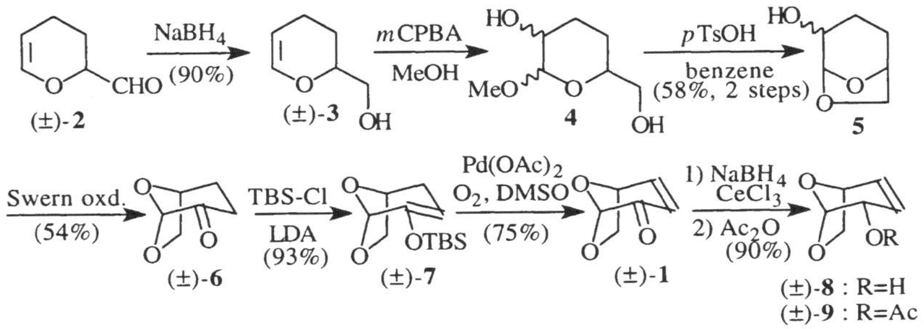

(±)-**2** → $NaBH_4$ (90%) → (±)-**3** → *m*CPBA, MeOH → **4** → *p*TsOH, benzene (58%, 2 steps) → **5** → Swern oxd. (54%) → (±)-**6** → TBS-Cl, LDA (93%) → (±)-**7** → $Pd(OAc)_2$, O_2, DMSO (75%) → (±)-**1** → 1) $NaBH_4$, $CeCl_3$ 2) Ac_2O (90%) → (±)-**8** : R=H; (±)-**9** : R=Ac

Scheme 1

On reaction with vinyl acetate in THF in the presence of immobilized lipase (lipase AK), (±)-**8** afforded the optically enriched acetate (+)-**9** (46%:70% ee), leaving the optically enriched alcohol (–)-**8** (46%:70% ee). On stirring in a phosphate buffer in the presence of lipase PS, (±)-**9** afforded enantiocomplementarily the acetate (–)-**9** (31%:96% ee) and the alcohol (+)-**8** (62%:74% ee). Optically enriched levoglucosenone **1** was obtained from the resolved products under standard conditions (**Scheme 2**).

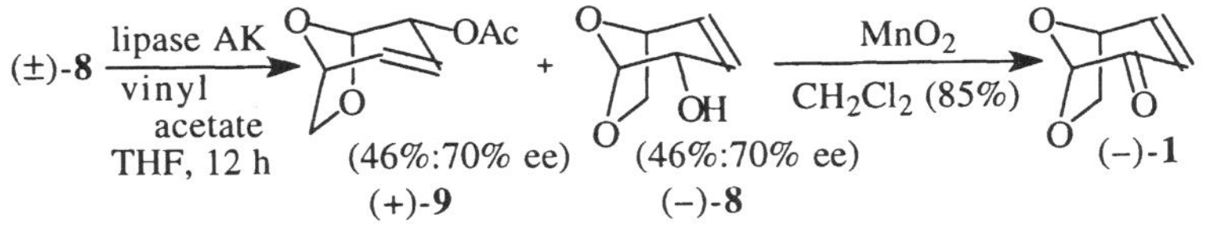

(±)-**8** → lipase AK, vinyl acetate, THF, 12 h → (+)-**9** (46%:70% ee) + (–)-**8** (46%:70% ee) → MnO_2, CH_2Cl_2 (85%) → (–)-**1**

Scheme 2

To obtain the enantiopure products, we explored an alternative procedure again by employing lipase-mediated resolution.[5] It is well known that a furfuryl alcohol furnishes a 3-pyrone hemiacetal on oxidative treatment.[6] Actually, the reaction of (2-furfuryl)ethylene glycol **13**, obtained[7] from furan **10**, with *m*CPBA afforded isolevoglucosenone[8,9] (±)-**15** having the opposite enone disposition to **1** after acid-cyclization of the pyrone **14**. For enzymatic resolution, (±)-**15** was converted into the alcohol (±)-**16** and the acetate (±)-**17**, diastereoselectively (**Scheme 3**).

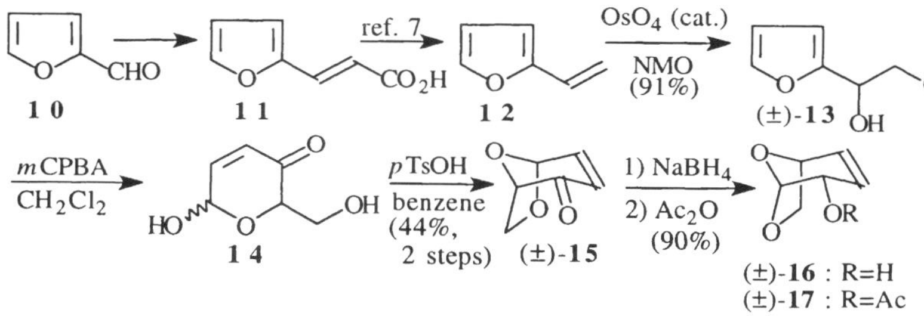

Scheme 3

Among the lipases examined, lipase AK gave the best result to furnish enantiopure (–)-**17** (48%:>99% ee), leaving enantiopure (+)-**16** (47%:>99% ee), when the racemate (±)-**16** was treated under transesterification conditions with vinyl acetate in THF. On the other hand, the racemate (±)-**17**, under hydrolysis conditions in the presence of lipase PS in a phosphate buffer, furnished enantiocomplementarily the optically enriched (–)-**16** (51%:97% ee), leaving the optically enriched (+)-**17** (48%:98% ee). The resolved products were converted into isolevoglucosenone **15** under standard conditions. Isolevoglucosenone **15** could be transformed into levoglucosenone **1** in three steps involving the Wharton rearrangement[10] *via* **18** and **19** (**Scheme 4**).

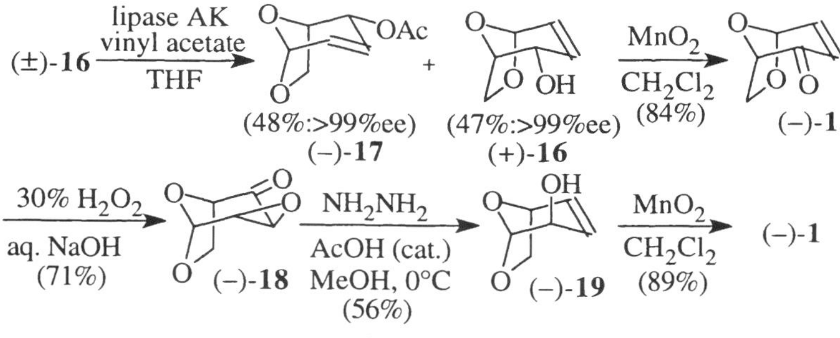

Scheme 4

Enantiocontrolled synthesis[10] of levoglucosenone **1** employing asymmetric dihydroxylation[11] (AD) has also been developed starting with 2-vinylfuran **12**. Thus, reaction of **12** with AD-mix-α afforded the optically enriched diol (+)-**13** (81%:90% ee) and with AD-mix-β afforded (–)-**13** (89%:93% ee). By employing the procedure above, optically enriched levoglucosenone **1** could be obtained *via* **15** (**Scheme 5**).

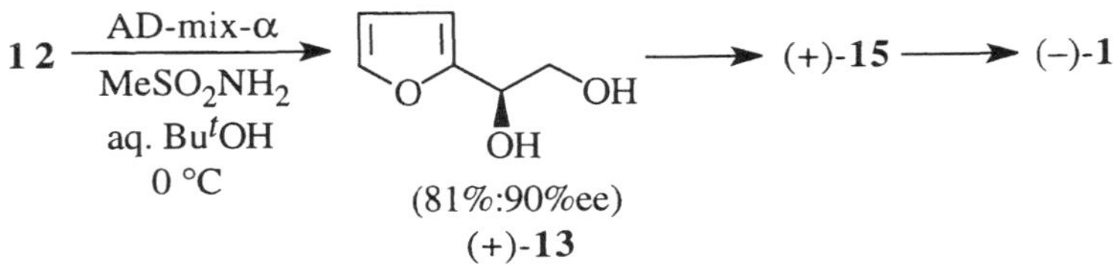

12 → AD-mix-α, $MeSO_2NH_2$, aq. Bu^tOH, 0 °C → (+)-13 (81%:90%ee) → (+)-15 → (–)-1

Scheme 5

Synthesis of a Levoglucosenone-Type Chiral Building Block

As noted above, one difficulty which prevents versatile use of levoglucosenone **1** is its sturdy internal acetal functionality. We, therefore, designed[12] a levoglucosenone carrying a handle on an appropriate position such as 6-alkoxymethyllevoglucosenone **20** so as to cleave the internal acetal functionality without difficulty. If **20** is available, its acetal linkage may be cleaved after conversion into a halomethyl derivative **21** to give a hemiacetal **22** under reductive conditions (**Scheme 6**).

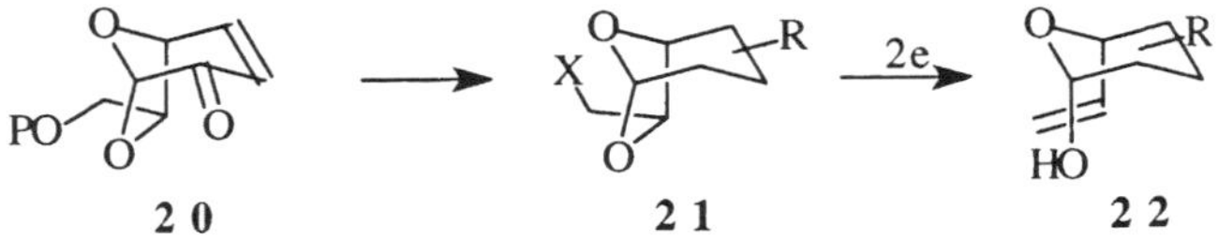

20 → 21 → (2e) → 22

Scheme 6

In order to realize this basic idea, we started with the *O*-protected 3(2-furyl)-2-propenol[13] **23** obtained from **11**. Thus, dihydroxylation of (±)-**23** gave the diol (±)-**24** which was transformed into 6-alkoxymethyl isolevoglucosenone (±)-**26** *via* **25** by sequential oxidative ring expansion and cyclization. (±)-**26** was then transformed into (±)-**27** and (±)-**28** for lipase-mediated resolution (**Scheme 7**).

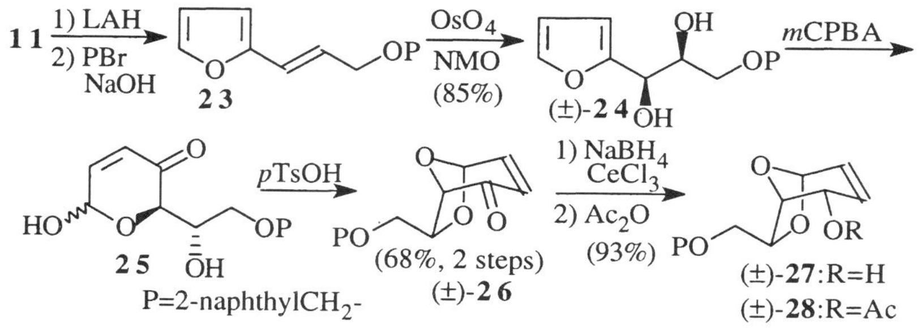

Scheme 7

On stirring with vinyl acetate in THF in the presence of lipase PS, (±)-**27** afforded enantiopure (–)-**28**, leaving enantiopure (+)-**27**. Interestingly, the attachment of an extra alkoxymethyl amplified the enantiodiscrimination in the enzymatic reaction.[13] The amplification was also observed when (±)-**28** was stirred in a phosphate buffer in the presence of the same lipase to afford enantiopure (–)-**27** and enantiopure (+)-**28**. The enantiopure products were reverted to the enantiopure isolevoglucosenone[13] **26** under standard conditions[15] (**Scheme 8**).

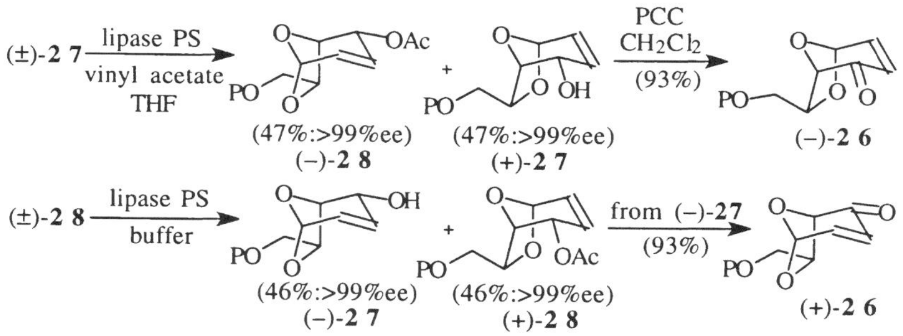

Scheme 8

Interestingly, again the alkoxymethyl attachment amplified the enantioselectivity in the AD reaction. Thus, **23**, on reaction with AD-mix-α reagent, afforded the enantiopure (+)-**24** while, with AD-mix-β reagent, it afforded the enantiopure (–)-**24**. Enantiopure products gave enantiopure **26** under standard conditions[12] (**Scheme 9**).

Isomerization of **26** to the levoglucosenone **20** was, however, found to be unexpectedly difficult under the Wharton conditions.[10] In contrast to **18**, **30** afforded only a minor amount of **31** on exposure to hydrazine. We had, therefore, to take a circuitous route to convert **26** into **20**. Thus, the *endo*-alcohol (+)-**27** obtained from **26**, was first inverted to

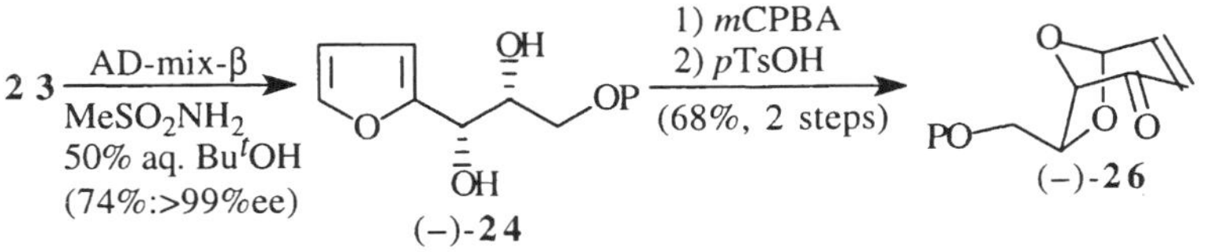

Scheme 9

the *exo*-alcohol **32** by the Mitsunobu reaction. On mesylation followed by solvolysis with aqueous calcium carbonate, **32** afforded the isomeric *exo*-alcohol **31**, by S_N2' substitution, which afforded **20** on oxidation. Overall yield of (+)-**20** from (–)-**26** was 33% in 6 steps[16] (**Scheme 10**).

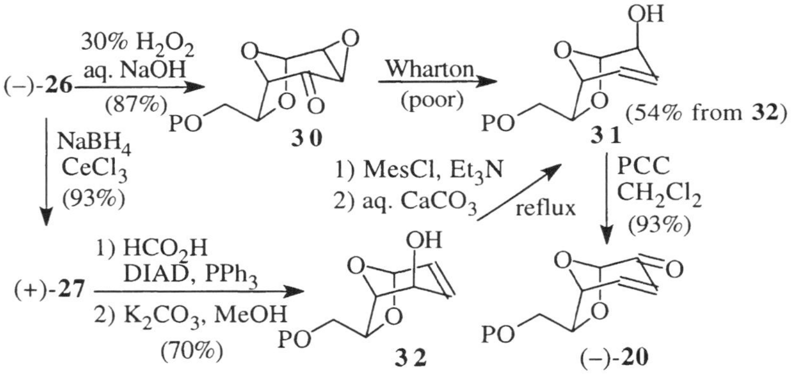

Scheme 10

Synthetic Exploitation of the Levoglucosenone Type Chiral Building Block

Having established the chiral synthesis of levoglucosenone **1** and its functionalized analogues in both enantiomeric forms, we next investigated the exploitation of **26**, the functionalized isolevoglucosenone, for the enantiocontrolled construction of natural products on the basis of its inherent convex-face selectivity and functionality, in particular, the alkoxymethyl handle for the acetal cleavage.

(a) Synthesis of the Eight L-Hexoses

As the most appropriate targets for demonstrating the potential of **26**, we chose the aldohexoses having eight diastereomers.[12,16] So far, only one method[17] has been developed for the synthesis of all of the eight possible hexoses by the groups of Masamune and Sharpless[18] who employed the asymmetric epoxidation (AE) as the key step. This procedure requires two AE steps and one carbon-carbon elongation step to obtain one particular hexose from a common four-carbon starting material. To develop a simpler method, we planned to converge all the eight hexoses into **26** through a diastereocontrolled introduction of three hydroxy functionalities on its enone moiety to generate the eight possible diastereomeric precursors corresponding to the eight targets.

Thus, **27**, generated from (–)-**26**, was transformed into **33** in 3 steps involving convex-face selective dihydroxylation. Conversion of **34** into the iodide **35** followed by treatment of the latter with zinc allowed facile cleavage of the acetal linkage to give **36** as expected. The generation of the hemiacetal **36** opened two ways leading to hexoses. Namely, on acetalization, followed by sequential oxidative cleavage and reduction of the vinyl functionality, **36** yielded **37** serving as the precursor of L-gulose (**route A**), while sequential reduction of the hemiacetal functionality and oxidative cleavage of the vinyl functionality yielded **38** serving as the precursor of D-glucose (**route B**). Moreover, **36** afforded the lactone **39** which was found to be epimerized to thermodynamically more stable **40**, on treatment with DABCO, serving as the precursor for L-idose. The demonstrated synthesis producing the three hexoses *via* the single hemiacetal **36** indicated that all of the eight trioxygenated precursors are not necessarily required for the production of the eight hexoses (**Scheme 11**).

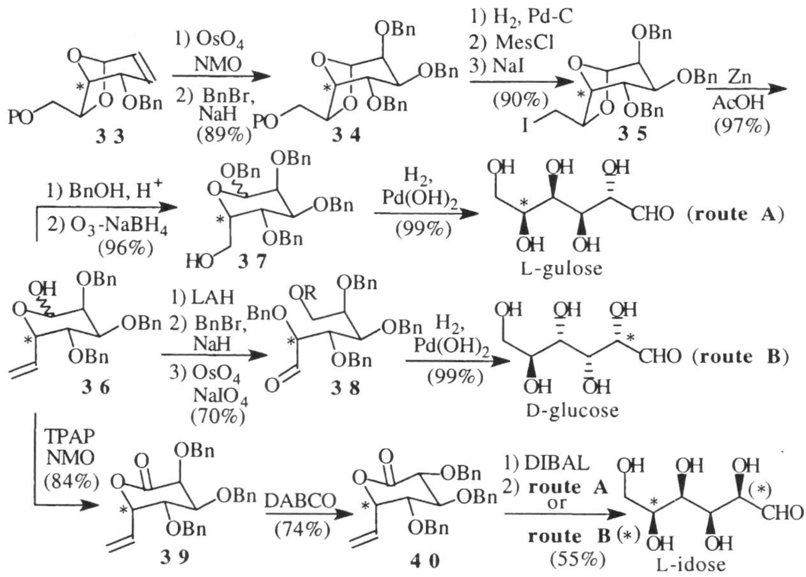

33 → 1) OsO_4, NMO; 2) BnBr, NaH (89%) → 34 → 1) H_2, Pd-C; 2) MesCl; 3) NaI (90%) → 35 → Zn, AcOH (97%) → 36

36 → 1) BnOH, H^+; 2) O_3-$NaBH_4$ (96%) → 37 → H_2, $Pd(OH)_2$ (99%) → L-gulose (route A)

36 → 1) LAH; 2) BnBr, NaH; 3) OsO_4, $NaIO_4$ (70%) → 38 → H_2, $Pd(OH)_2$ (99%) → D-glucose (route B)

36 → TPAP, NMO (84%) → 39 → DABCO (74%) → 40 → 1) DIBAL; 2) route A or route B (*) (55%) → L-idose

Scheme 11

To obtain other diastereomeric hexoses, **30**, obtained from (–)-**26**, was first converted into the benzoate **41**. On exposure to boron trifluoride,[19] **41** furnished **43**, through **42**, after methanolysis. Employing the same procedure above, **43** was converted to the hemiacetal **44** which furnished L-galactose *via* **route A** and D-galactose *via* **route B** (**Scheme 12**).[12]

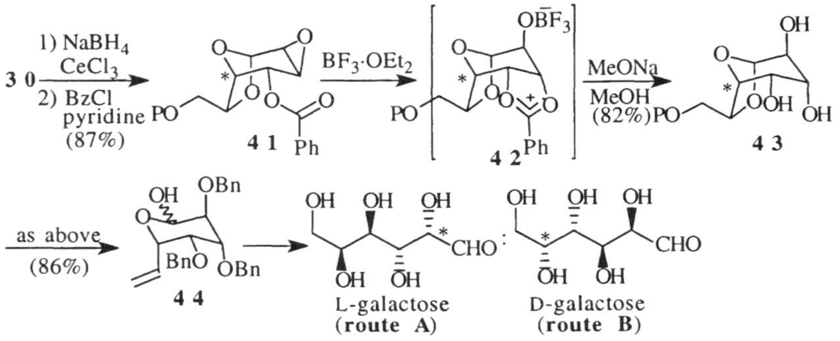

Scheme 12

On the other hand, **32** was first transformed into the 3,5-dinitrobenzoate **45** which allowed diastereoselective dihydroxylation[20] to give **46** after methanolysis and benzylation. **46** afforded the hemiacetal **47** from which L-allose *via* **route A** and D-allose *via* **route B** were obtained (**Scheme 13**).[16]

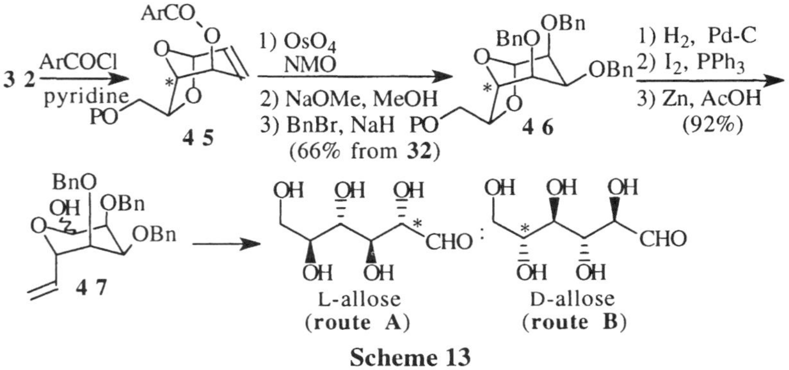

Scheme 13

The benzyl ether **48**, generated from (+)-**20**, afforded **49** which furnished L-altrose *via* **route A** and L-talose *via* **route B** through the hemiacetal **50** (**Scheme 14**).[16]

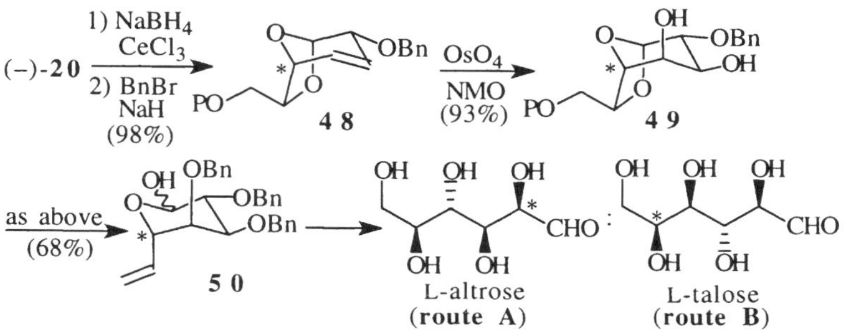

Scheme 14

The epoxide **51**, obtained from (+)-**20**, afforded **52** from which L-mannose was obtained by either **route A** or **route B** through the hemiacetal **53**. Moreover, **53** furnished L-glucose *via* **route A** and D-gulose by **route B** after conversion into **54** and **55** (**Scheme 15**).[16]

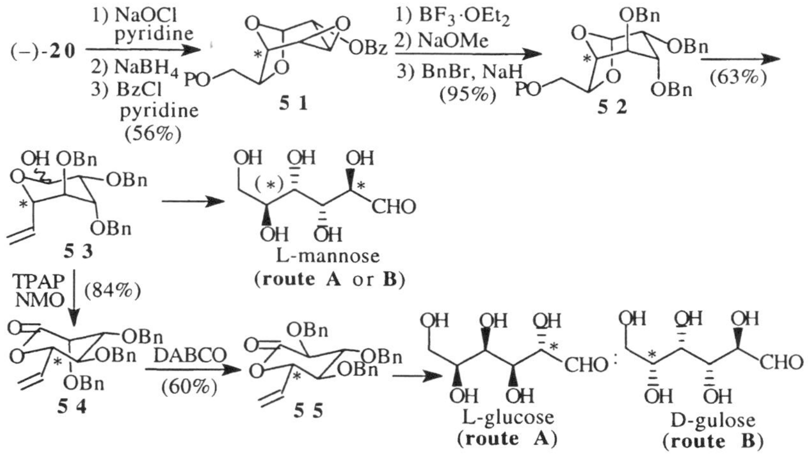

Scheme 15

Thus, the eight L-hexoses were obtained from the single precursor (–)-**26** along with additional four D-hexoses through the five hemiacetal intermediates, **36**, **44**, **47**, **50** and **53**.

(b) Synthesis of L-Novioses

Inherent convex-face selectivity and functionality of **26** enable us to construct other sugar molecules as well as some other natural products.

Starting with (+)-**26**, L-noviose, the sugar moiety of antibiotic novobiocin, was obtained. Thus, (+)-**26** was converted into **56**, *via* **41**, which on reaction with methyllithium followed by oxidative cleavage and deprotection gave L-noviose (**Scheme 16**).[21]

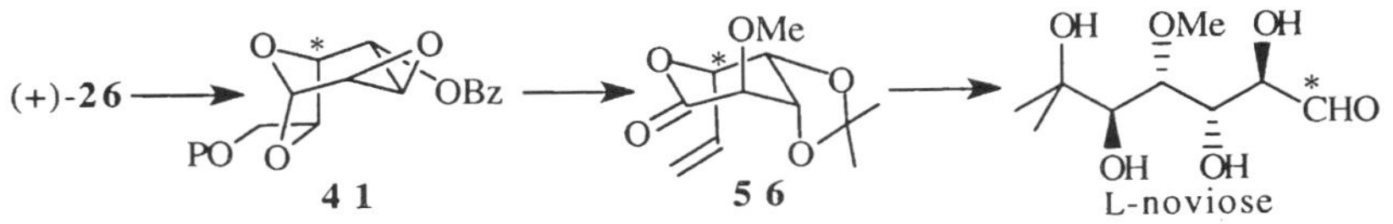

Scheme 16

(c) Synthesis of (+)-Conduritol F and (+)-Febrifugine

The terminal olefin functionality generated by the cleavage of the acetal linkage was conveniently used in the ring-closing metathesis[22] (RCM) which led to a cyclitol (+)-conduritol F, an antimalarial (+)-febrifudine, the C_{28}-C_{34} segment of an immunosuppressant FK-506, and a key biosynthetic precursor (–)-shikimic acid.

Thus, (–)-**26** was transformed into the diene **58** which, on sequential RCM and deprotection, furnished (+)-conduritol F (**Scheme 17**).[23] On the other hand, (–)-**26** was converted into the diene **60**, *via* **59**, which gave the ketone **62**, the precursor of febrifugine,[24] by conversion involving RCM.

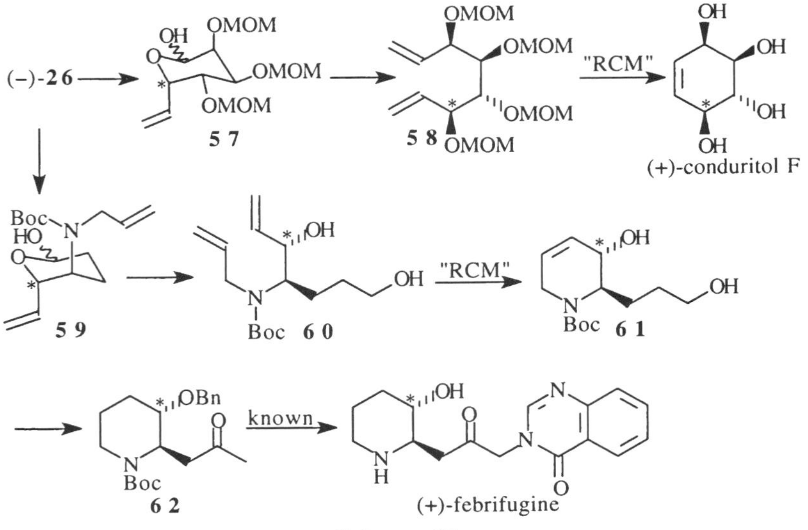

Scheme 17

(d) Synthesis of the C_{28}-C_{34} Fragment of FK-506 and (–)-Shikimic Acid

The C_{28}-C_{34} fragment of FK-506 and (–)-shikimic acid required the hemiacetal **63** having the same framework which was obtained from (+)-**26** by convex-face 1,4-addition and acetal cleavage. The C_{28}-C_{34} fragment was obtained through **63~65** (R=Me) on sequential RCM and hydrogenation.[25]

On the other hand, **65** (R=MOM), obtained through **63~65** (R=MOM) by employing the same sequence involving RCM, was converted into **66** on sequential diastereoselective epoxidation, protection and oxidation. On treatment with DBU, **66** afforded **67** serving as the precursor of (–)-shikimic acid (**Scheme 18**).[26]

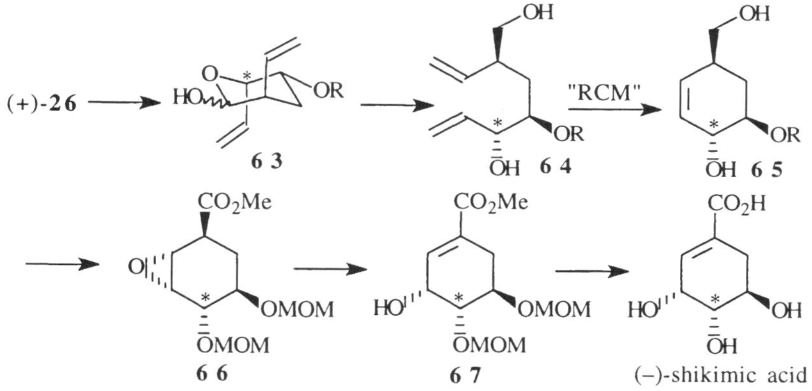

(+)-26 → 63 → 64 "RCM" → 65 → 66 → 67 → (–)-shikimic acid

Scheme 18

(e) Synthesis of (–)-Kainic Acid

Combination of **26** with pericyclic reaction led to a diastereocontrolled synthesis of anthelmintic (–)-kainic acid and anticholinergic (–)-physostigmine and (–)-physovenine. Thermolysis of **68**, obtained from (+)-**26** furnished diastereoselectively **69** by intramolecular ene reaction. (–)-Kainic acid was obtained from **69** through a sequence of several steps of reactions (**Scheme 19**).[27]

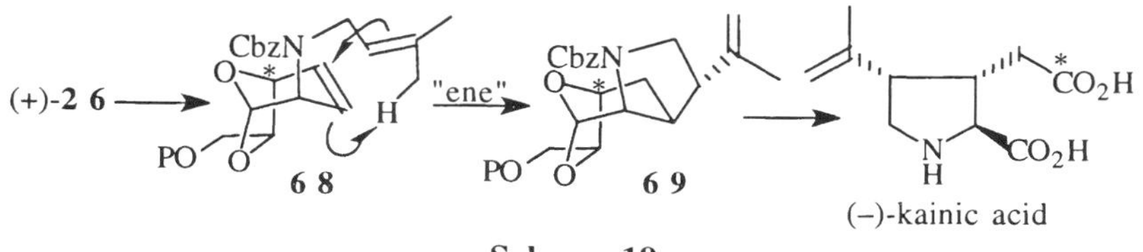

(+)-26 → 68 "ene" → 69 → (–)-kainic acid

Scheme 19

(f) Synthesis of (–)-Physostigmine and (–)-Physovenine

The Fischer indolization of **70** obtained from (–)-**26**, with arylhydrazine proceeded diastereoselectively to give rise to the indolenine **72** *via* **71**. Both (–)-physostigmine and (–)-physovenine were obtained from **72** through a sequence of several steps of reactions (**Scheme 20**).[28]

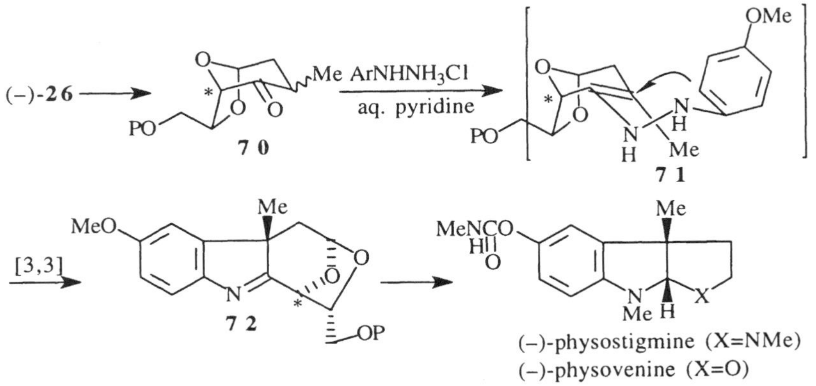

Scheme 20

Conclusion

As demonstrated, we now have developed a second method being capable of producing all of the eight hexoses from a single artificial chiral building block obtained by employing either enzymatic resolution or asymmetric synthesis. On the basis of the biased structure and the functionality involved in the newly designed levoglucosenone block, not only the eight possible hexoses, but other biologically interesting natural products could also be constructed.

Acknowledgments

I express my gratitude to my coworkers who have participated in the work herein reviewed whose names appear in the references. Our work cited in this account has been largely supported by grants from the Ministry of Education, Science, Sports and Culture, Japan.

References

1. a) Witczak, Z. J. *Synlett* **1996**, 108 and references cited therein.

2. Ebata, T.; Matsushita, H. *J. Syn. Org. Chem. Jpn.* **1994**, *52*, 1074.
3. Shibagaki, M.; Takahashi, K.; Kuno, H.; Honda, H.; Matushita, H. *Chem. Lett.* **1990**, 307.
4. Unpublished result.
5. Unpublished result.
6. Achmatowicz, O. "Organic Synthesis Today and Tomorrow" Trost. B. M.; Hutchinson, C. R. Ed. Pergamon, Oxford 1980, pp 307-325.
7. Schmidt, U.; Werner, J. *Synthesis* **1986**, 986.
8. Köll, P.; Schultek, T.; Rennecke, R. -W. *Chem. Ber.* **1976**, 337.
9. Horton, D.; Roski, J. P.; Norris, P. *J. Org. Chem.* **1996**, *61*, 3783.
10. Taniguchi, T.; Nakamura, K.; Ogasawara, K. *Synlett* **1996**, 971.
11. Kolb, H. C.; VanNieuwenhze, M. S.; Sharpless, K. B. *Chem. Rev.* **1994**, *94*, 2483.
12. Takeuchi, M.; Taniguchi, T.; Ogasawara, K. *Synthesis* **1999**, 341.
13. Three protecting groups (NAP: 2-naphthylmethyl[14], Bn: benzyl and TBS: *tert*butyldimethylsilyl) were employed. Enantioselectivity was virtually not dependent on the protecting groups employed.
14. Gaunt, M. J.; Yu, J.; Spencer, J. B. *J. Org. Chem.* **1998**, *63*, 4172.
15. Taniguchi, T.; Takeuchi, M.; Kadota, K.; ElAzab, A. S.; Ogasawara, K. *Synthesis* **1999**, 1325.
16. Takeuchi, M.; Taniguchi, M.; Ogasawara, K. *Chirality* in press.
17. McGarvey, G. J.; Kimura, M.; Oh, T.; Williams, J. M. *Carbohydrate Chem.* **1984**, *3*, 125.
18. (a) Ko, S. Y.; Lee, A. W. M.; Masamune, S.; Reed, L. A., III; Sharpless, K. B.; Walker, F. J. *Science* **1983**, *220*, 949. (b) Ko, S. Y.; Lee, A. W. M.; Masamune, S.; Reed, L. A., III; Sharpless, K. B.; Walker, F. J. *Tetrahedron* **1990**, *46*, 245.
19. Prystas, M.; Gustafsson, H.; Sorm, F. Coll. Czech. *Chem. Commun.* **1971**, *36*, 1487.
20. Matsumoto, K.; Ebata, T.; Koseki, K.; Kawakami, H.; Matsushita, H. *Heterocycles* **1991**, *32*, 2225.
21. Takeuchi, M.; Taniguchi, T.; Ogasawara, K. *Tetrahedron Lett.* in press.
22. Grubbs, R. H.; Chang, S. *Tetrahedron* **1998**, *54*, 4413.
23. Unpublished result.
24. Unpublished result.
25. Takeuchi, M.; Taniguchi, T.; Ogasawara, K. *Tetrahedron: Asymmetry*, in press.
26. Unpublished result.
27. Unpublished result.
28. Unpublished result.

Chapter 4

Sugar-Derived Building Blocks for the Synthesis of Non-Carbohydrate Natural Products

Frieder W. Lichtenthaler

Institute of Organic Chemistry, Darmstadt University of Technology, D–64287 Darmstadt, Germany

The generation of enantiopure non-carbohydrate natural products from readily available sugars is of *practical value* only, if the individual reactions employed allow simple reagents, proceed uniformly, and avoid complex separations in work-up procedures to ultimately enable favorable overall yields. Such practical criteria entail the transformation of a sugar, "overfunctionalized" with chirality and hydroxyl groups, into an enantiopure building block with suitable functionalities, e.g. C=C or C=O groups, or both, and only one or two centers of asymmetry left. The principal possibilities for the elaboration of new benign *reaction channels* sugar → enantiopure building block are delineated with an emphasis on dihydropyranones with isolated or vicinal centers of chirality, as they provide a more clearly foreseeable stereochemistry in additions of *O*-, *N*- and *C*-nucleophiles than their open-chain or furanoid counterparts. Application of various of such hexose-derived six-carbon building blocks to the synthesis of diplodialide-type pheromones, of the soft coral metabolites (*S,S*)-palythazin and (*S,S*)-bissetone, of ACRL-Toxin, of a series of *Labiatea*-derived C_{12}-enelactones, and of uscharidine-type cardenolides is presented.

Carbohydrates are the single most abundant class of organic compounds associated with living matter, and, hence, are enantiopure. This auspicious fact together with the bulk-scale availability at low cost renders them ideal starting materials for organic preparative purposes if the acquisition of an enantiomerically homogeneous target molecule is a conditio sine qua non – an approach that is a most attractive alternative to the construction of enantiopure target molecules by asymmetric synthesis.

Despite of these highly favorable prerequisites of carbohydrates in general, and low-molecular weight sugars in particular – Table 1 gives an overview on their accessibility – it appears surprising that sugars are not utilized on a much larger scale as raw materials for chemical industry (*1-7*) on one hand, and for the construction of enantiopure non-carbohydrate natural products and pharmaceuticals on the other. There are reasons for this, of course. Sucrose, for example, "the royal carbohydrate" (*8*), and with an annual generation of 130 million tons the world's most abundantly produced organic compound, provides an interesting chemisty (*9,10*), yet is unsuited for many synthetic transformations due to its acid-sensitive intersaccharide linkage. Its component monosaccharides, D-glucose and D-fructose, are devoid of this deficiency, yet direct utilization of their vast synthetic potential is impeded by a number of obstacles: they are overfunctionalized with hydroxyl groups of similar or identical reactivities, they have considerably more chiral centers along the six-carbon chain than required for non-sugar target molecules, and, they lack suitable functional groups such as olefinic or carbonyl unsaturation to which modern organic preparative methodology can directly be applied.

MONOSACCHARIDES AS ENANTIOPURE EDUCTS

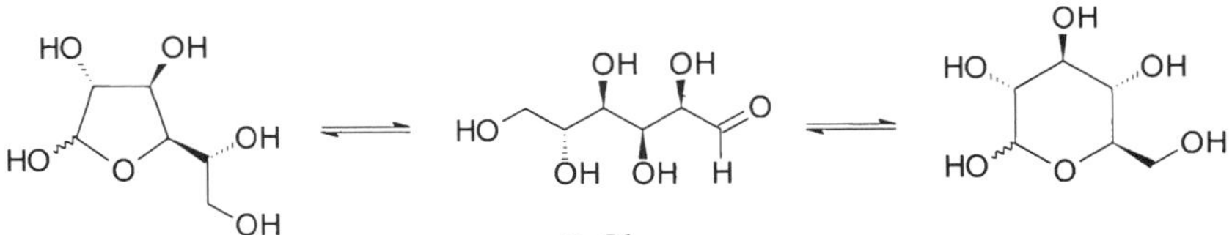

D-Glucose

SHORTCOMINGS: overfunctionalized with hydroxyl groups
too many chirality centres
lack of C=C and C=O functionalities

Table 1. Annual Production Volume and Prices of Simple Sugars, Sugar-derived Alcohols and Acids.

		World production [a] (metric t/year)	Price [b] (€/kg)
Sugars	Sucrose	130.000.000	0.30
	D-Glucose	5.000.000	0.60
	Lactose	295.000	0.60
	D-Fructose	60.000	1.00
	Isomaltulose	50.000	2.00
	Maltose	3.000	3.00
	D-Xylose	25.000	4.50
	L-Sorbose	60.000	7.50
Sugar Alcohols	D-Sorbitol	650.000	1.80
	D-Xylitol	30.000	5.00
	D-Mannitol	30.000	8.00
Sugar-derived Acids	D-Gluconic acid	60.000	1.40
	L-Lactic acid	> 100.000	1.75
	Citric acid	500.000	2.50
	L-Tartaric acid	35.000	6.00
Amino Acids	L-Lysine	40.000	5.50
	L-Glutamic acid	500.000	7.00

[a] Reliable data are only available for the world production of sucrose, the figure given referring to the crop cycle 2000/2001 (*11*). All other data are average values based on estimates from producers and/or suppliers, as the production volume of many products is not publicly available. [b] Prices given are those attainable in early 2001 for bulk delivery of crystalline material (where applicable) based on pricing information from sugar industry. The listings are intended as a benchmark rather than as a basis for negotiations between producers and customers. Quotations for less pure products are, in part, sizeably lower, e.g. for the commercial sweetener "high fructose syrup", which contains up to 95% fructose, and, thus, may readily be used for large-scale preparative purposes.

These adverse conditions have elicited considerable efforts to reduce the number of chiral centers as well as hydroxyl groups with the simultaneous introduction of useful functional groups (*10*, *13-15*). One approach involves the shortening of the aldose carbon chain, or, more simply, its bisection, as exemplified by the use of D-mannitol-derived 2,3-*O*-isopropylidene-D-glyceraldehyde. Whilst this product and its L-ascorbic acid-derived enantiomer have developed into popular enantiopure three-carbon synthons (*16*), it may be objected that the photosynthetic achievement of Nature which graciously provides us with six-carbon compounds, is utilized rather inefficiently, clearly pointing towards elaboration of synthons from sugars *with* retention of the carbon chain.

Indeed, the most frequently used alternative to sugar-derived three-carbon synthons is the gradual step-by-step carving out of a target molecule of a segment thereof; from a hexose, resulting in a reaction sequence that is specifically tailor-made for the synthetic target. The number of complex, non-carbohydrate natural products synthesized via this approach is enormous (*12*, *17-19*). The vast majority of these total syntheses, however, are exceptionally long and cumbersome, and their transposition to a reasonably large scale is essentially unfeasible with respect to the reagents used, the number of steps required, the expenditure of work involved, and the overall yields attainable.

Thus to fully exploit the huge potential lying in the readily available pentoses and hexoses, criteria of efficiency, practicality environmental benignity, and overall economy have to be applied to the ensuing reactions to be performed – not the least in the anticipation that the process evolving may eventually be used industrially. Such criteria obviously comprise

- retention of the carbon-chain of the sugar,
- selection of reactions that allow for simple reagents a uniform course, and an uncomplicated, non-chromatographic workup,
- use of simple protecting groups, if not avoidable at all,
- steering for stable, crystalline, readily purificable intermediate products along the way,
- reasonably high overall yields, i. e. 75 % per step on the average,
- overall reaction sequences that have the potential of being transposable into the hectogram scale.

Realization of most or all of these criteria calls for the conversion of a pentose or hexose into a versatile five- or six-carbon synthon, preferably a stable building block with one or two chiral centers and with synthetically flexible functional groups. Since the efficiency of this conversion largely determines the practical value of the total synthesis to be accomplished, restriction is necessary to what is preparatively "makeable" in 4-5 steps and with overall yields of 40-50 %. This, in turn, reduces the number of methodical entries sugar ⇒ building block to a very few "reaction channels" which, nolens volens, are different for each sugar, since an optimal compliance with their individual stereochemical intricacies is imperative for achieving adequate preparative results.

The initial stage of any *reaction channel* from a sugar to an enantiopure building block invariably involves fixation of the sugar in the respective tautomeric form. The few preparatively useful reactions of this type have been elaborated long ago, usually dating back to the beginning of carbohydrate chemistry over a 100 years ago: mercaptalization to the **acyclic** dithio acetals (*17*), isopropylidenation to **furanoid** systems – as e. g. the preparation of "diacetone glucose" by Emil Fischer in 1885 (*18*), or the generation of **pyranoid** structures, such as glycosides (*19*), glycal (*20*) or hydroxyglycal esters (*21*):

Diacetonide

Dithioacetal

Me_2CO / $ZnCl_2$ 90 %

70 % EtSH / H^+

D-Glucose

70 % MeOH H^+

70 % 1. Ac_2O 2. Br_2 / P 3. Zn / HOAc

85 % 1. BzCl 2. Br_2 / P 3. Et_2NH

Glucoside

Glucal

Hydroxyglucal

Once, tautomeric fixation has been achieved, the mono- or disaccharide is to be converted into building blocks with useful functionality, such that the modern preparative armoury of organic chemistry can directly be applied. Thereby, the number of furanoid, open-chain, and pyranoid building blocks is immense, in principle, yet when imposing the practical norms outlined for their acquisition, the possibilities shrink substantially, as there are comparatively few preparatively satisfactory methodologies available, *reaction channels* so to say, and their outcome is usually dependent on the inherent stereochemistry of the individual sugar derivative, such that transfer of reaction conditions from one sugar to another rarely results in analogous products.

To be presented in the sequel, are a series of prototype *reaction channels* leading from simple, tautomerically fixed D-glucose derivatives to enantiopure building blocks along preparatively useful, practical protocols, followed by their utilization for the straightforward total synthesis of a series of natural and non-natural products in optically active, enantiomerically homogeneous form.

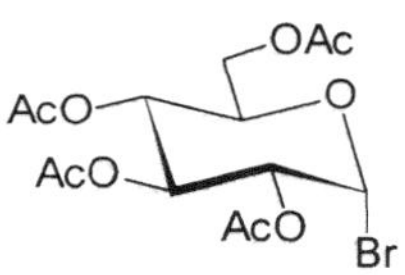

1	**2**	**3**
	R = Ac, Bz, Bn	
85 % (2 steps)	70 - 75 % (4 steps)	50 % (5 steps)

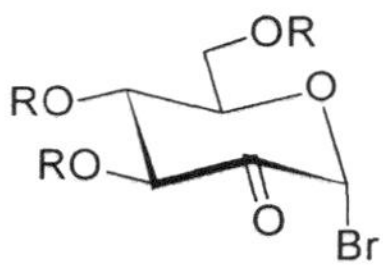

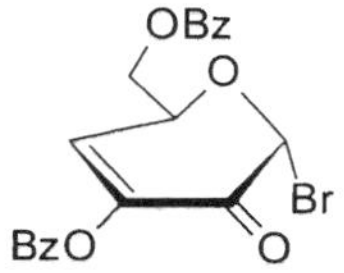

By contrast, 2-oxoglycosyl bromides of type **2** (" ulosyl bromides "), with an electron-withdrawing carbonyl function at C-2 rather than a participating acyloxy substituent as in **1**, have only recently become well accessible (*25*, *26*). Similarly, the original protocol for the obtention of enolone bromide **3** (*27*) – an "advanced" building block due to only two asymmetric centers which flank the versatile enolone ester functionality – has only recently been optimized (*28*) to allow overall yields from D-glucose in the 55 % range.

Acylated 2-oxoglycosyl bromides of type II may efficiently be generated from hydroxyglycal esters by either of two ways, i. e. a high-yield, three-step procedure involving hydroxylaminolysis (*29*), deoximation (*30*), and photobromination (*31*), or, alternately, by a one-step process, simply consisting of exposure of hydroxyglycal ester I, in dichloromethane solution, to NBS or bromine in the presence of methanol (*25*, *26*). Mechanistically, the direct conversion I → II is thought to proceed via initial attack of a brominium ion to a 2-bromobenzoxonium salt intermediate of type III, in which the 2-*O*-benzoyl group is captured by methanol; the resulting formation of methyl benzoate leaves ion pair IV that combines to II. The ease with which this conversion can be effected (30 min, room temperature) is as remarkable as the yields attainable (80-90 %) and the applicability to disaccharide-derived hydroxyglycal esters (*32*). Accordingly, glycos-2-ulosyl bromides of type II are nearly as well accessible from basic monosaccharides as the standard acylated glycosyl halides.

R	%
H	78
CH_3	83
$BzOCH_2$	89

Glycosidation of these ulosyl bromides under Koenigs-Knorr conditions proceeds in an essentially stereospecific manner, i. e. **2** → **4**, so that with a large variety of alcohols, even comparatively non-reactive saccharide OH groups, high yields of the respective β-ulosides are obtained. As the subsequent hydride reduction **4** → **5** proceeds with *manno*-selectivities of 5:1 to > 20:1, this methodology, which has been termed the "ulosyl bromide approach", has proved highly advantageous for the generation of β-D-mannose-containing oligosaccharides (*33*, *34*).

Anomeric C-homologation can be effected by applying Reformatsky conditions, zinc promoted addition to aldehydes smoothly providing C-glycosiduloses of type **6** (*35*, *36*). The synthetic potential of ulosyl bromides is similarly accentuated by their glycosidation with vicinal diols, ethylene glycol, for example, leading not only to the β-glycosidulose but by subsequent elaboration of a cyclohemiketal to pyranodioxanes of type **7** (*25*, *37*) – dioxane-anellated glycosides that occur in a variety of *Calotropa* cardenolides which may advantageously be synthesized for the first time by this methodology.

ROH / Ag_2CO_3 ~ 90 %

Ag_2CO_3 86 %

H⁺

Zn / RCHO

4 2 7

5 6 7

β-D-Mannoside C-Glycoside Pyranodioxane

In this context, pyranoid sugar enolones of type **8**, or **9**, i. e. those that carry chiral centers on either side of the enolone structural element, are even more powerful building blocks stereoselectivities in addition reactions. They are accessible in various substituted forms from the respective hydroxyglycal esters in another, preparatively delightful reaction channel, a chlorination → hydrolysis → elimination sequence (*27*, *28*).

1. Cl_2 / H_2O
2. $NaHCO_3$
79 %

8 9

In the case of D-glucose-derived hydroxyglycal ester **8**, low temperature chlorination affords an approximate 1:8-mixture of the *cis*-dichloride of α-D-*manno*-configuration (**10**) and the α-D-*gluco*-benzoxonium salt (**11**), of which the latter, on in situ hydrolysis with water, is converted into the α-D-hexulose **12**. Brief warming with moist $NaHCO_3$, or stirring with sodium acetate in acetone, induces β-elimination of benzoic acid in **12** to yield an 8:1 mixture of dihydropyranone **9** and the dichloride **10**, from which **9** can be isolated in pure form by a single crystallization in satisfactory overall yield (79 %) (*27*, *28*).

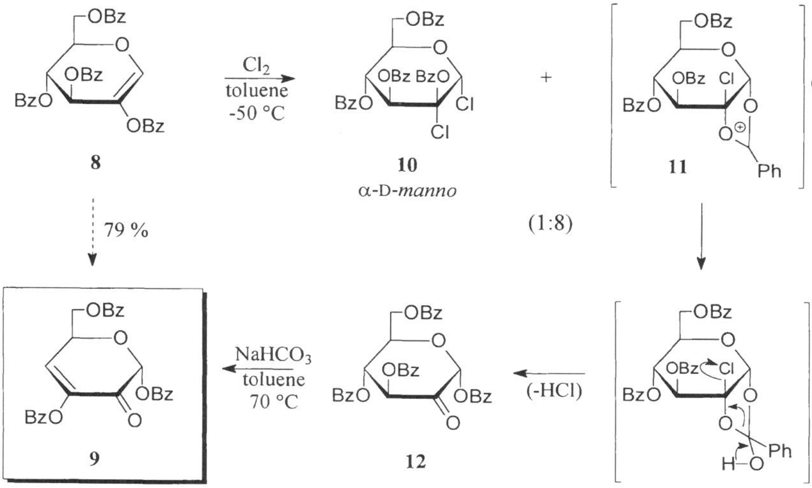

Thus, from the stage of a hydroxyglycal ester of type **8**, which is only a one-pot reaction away from a bulk sugar, another simple, practical one-pot reaction leads to enantiopure 2,6-dihydropyranones, which provide a prolific ensuing chemistry, particularly with respect to hydride addition (*38*), C-branching with Grignard or cuprate reagents (*39*), and Diels-Alder type cycloadditions (*40*). *O*-Nucleophiles attack at the carbonyl function from the sterically as well as electronically more favored pro-axial face to provide upon the usually ensuing benzoyl group migration tetrahydropyranones of type **13** which are bis-acetal derivatives of actinospectose in the form present in natural products of the spectinomycin type (cf. below). An even closer resemblance to dioxane-anellated glycosides exhibits the pyrano[2,3-*b*] dioxane **16** preparable from **9** in two high-yielding steps; being remarkably insensitive towards acidic conditions, it is easily converted into the respective 1-halides, as, for example, into the α-bromide **3** on treatment with HBr / acetic acid (*27*, *28*). Alcoholysis of such enolone bromides may be performed with high β-selectivity when employing sodium hydrogen carbonate or silver carbonate as acid scavenger, to give the β-glycosides in yields well over 80 % (*27*). Similarly, when subjected to glycolysis with ethylene glycol in the presence of Ag_2CO_3, bromide **3** smoothly elaborates the dioxane ring-anellated pyranone **16**, the β-selective glycosylation being followd by cyclization to the hemiketal and subsequent benzoyl migration (arrows in **15**) with liberation of the carbonyl function (*37*).

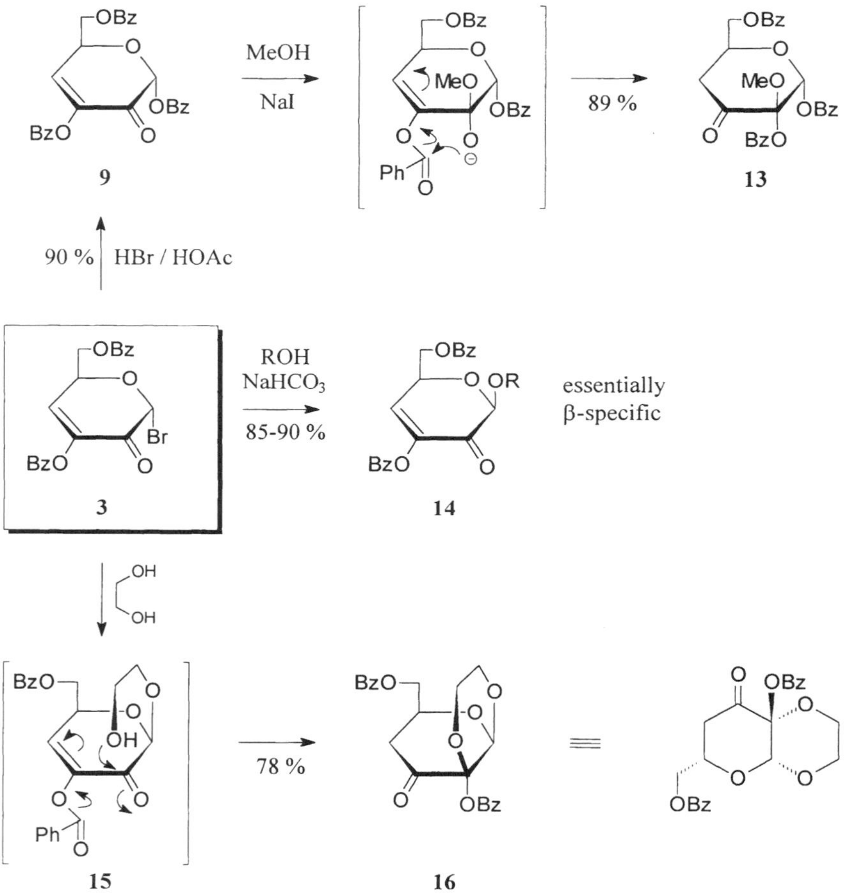

This "doubly glycosidic" anellation of – formally – a 2,3-diketosugar onto a diol is highly reminiscent structurally of the broad spectrum antibiotic spectinomycin (*41*) and the cardiac glycosides isolable from the latex and leaves of *Calotropis procera* (*42*) a bushy plant indigenious in wide parts of Africa and India. In these, a pyranoid 4-deoxy-2,3-dicarbonyl sugar, which has been designated actinospectose yet could only be "characterized" as an intractable tar (*41*), is fused to cyclohexanediol-type aglycons:

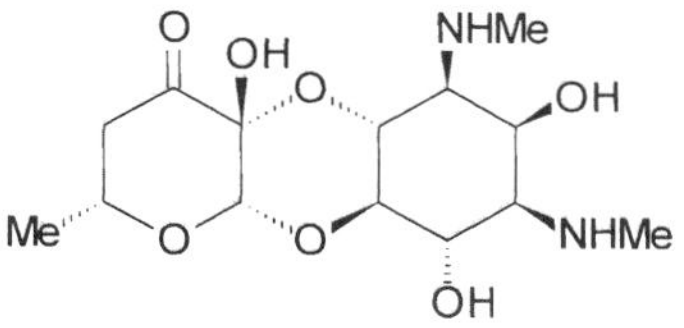

Spectinomycin

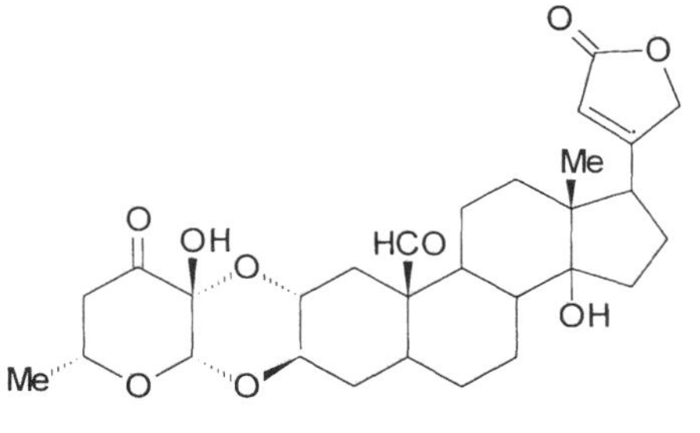

Uscharidin

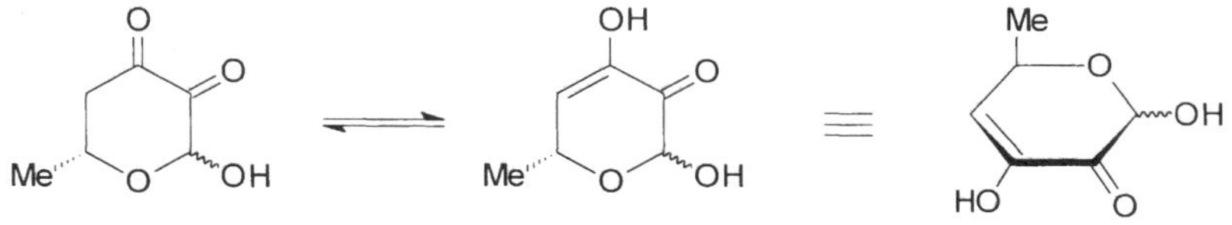

Sugar component: Actinospectose

Indeed, the insight into the intricate stereochemical details of ensuing reactions of the pyranoid enolone ester bromide **3** – the pyranoid enolone ester **14**, de facto, is an actinospectoside derivative, **16** a structural element of the spectinomycin as well as uscharidine-type cardiac glycosides (except for the terminal benzoyloxy group) – eventually led to practical synthetic strategies for their acquisition.

The *R*-methyl group in the pyranoid portion of spectinomycin pointed towards a 6-deoxy-hexose as the chiral educt and – given the ready elaboration of dihydropyranones of type from hydroxyglycal esters, outlined above – for an acylated 6-deoxy-D-hydroxyglucal as the actual starting material. The tribenzoate **17** was chosen for this purpose due to its expeditious preparation from methyl α-D-glucopyranoside in five large-scale adaptable, high yielding steps (*43*). When sequentially subjected to low temperature-chlorination, hydrolysis, and elimination – in a manner analogous to the conversion – and treatment with HCl, the crystalline actinospectosyl chloride **18** is readily obtained in acceptable overall yield (*43*).

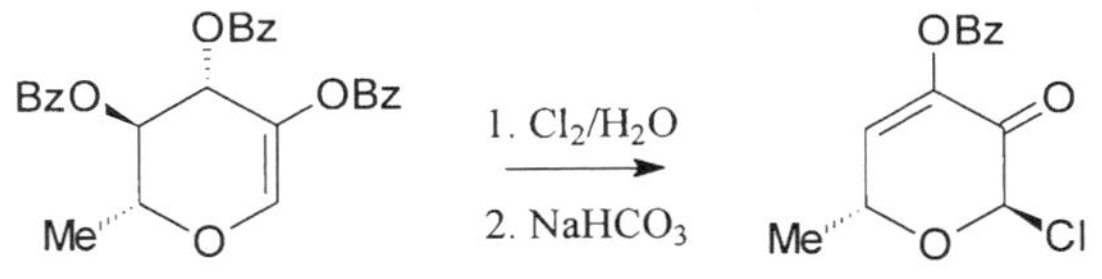

Exposing **18** and bis-carbobenzoxy-actinamine (**19**) to the usual glycosylation conditions (Ag_2CO_3), however, no reaction took place at room temperature, and at 50 °C it was sluggish and comparatively unselective, obviously due to the low reactivity of the actinamine-5-OH. A more forcing catalyst such as silver triflate, however, induced the desired β-selective glycosylation of the sterically less hindered 5-OH, whereafter ketalization and benzoyl group migration – in a manner analogous to (cf. above) – yielded the bis-carbobenzoxy-spectinomycin benzoate **21** in good yield (*37*) considering the three concomitant steps involved. Since de-*O*-benzoylation may readily be effected (K_2CO_3 / methanol) and hydrogenolysis is proceeding smoothly, this sequence constitutes a most facile, efficacious total synthesis of spectinomycin, requiring 12 steps from D-glucose with an overall yield of 9.9 %, averaging 80 % per step (*37*).

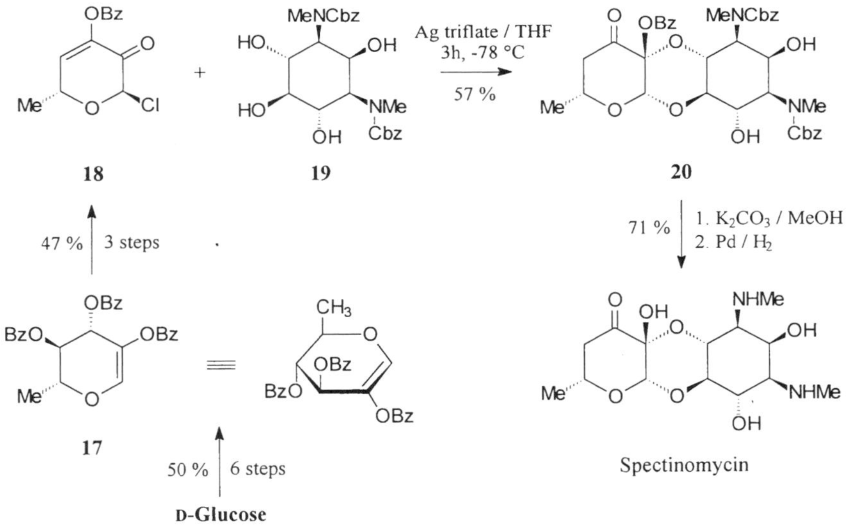

Spectinomycin

This expeditious route to spectinomycin not only compares favorably with two previous syntheses of this antibiotic, one elaborating the pyranoid half from D-glucose in over 20 steps (*44*), the other requiring nine from the rather incommodiously accessible L-glucose (*45*), in addition, this approach is versatile enough to lend itself to the synthesis of uscharidine type cardiac glycosides from steroidal diols.

Since calotropagenine, the aglycone of uscharidine, is accessible only with difficulty, model experiments have been carried out with the more readily available cholestan-2α,3β-diol **21**. On Ag_2CO_3-mediated reaction with actino-spectosyl chloride **18** two isomers formed smoothly in an approximate 3:1 ratio, of which the major one – as evidenced by ^{1}H- and ^{13}C-NMR data, corroborated by NOE experiments – proved to be the "unnatural" anellation product **23**. The minor product **22**, however, could be readily debenzoylated by treatment with butylammonium acetate in aqueous acetonitrile to afford the uscharidine analog **24** albeit in modest yield (*46*), yet crystalline form so that its linkage geometry could be secured by an X-ray structure analysis (*47*).

OBz, O, Me, O, Cl, HO, HO

18 + **21**

70 % Ag_2CO_3/CH_2Cl_2 16 h, 25 °C

BzO, OH, O, Me, O, O, Me, O, O, O, BzO, OH

22 + **23**

90 % Bu_4NOAc $MeCN/H_2O$

BzO, OH, O, Me, O, O

mp 142 - 144 °C

$[\alpha]_D^{20}$ - 15.5° (CHCl$_3$

24

The 3:1 glycosylation selectivity observed clearly points towards a higher reactivity of the 2-OH group in the diol educt **21** – a finding that similarly should enable the effective introduction of an O^2-blocking group, whereafter reactions with actinospectosyl chloride **18** will uniformly be directed to the "natural" glycosylation site. This obvious preparative solution has already been carried out with 2-*O*-benzyl derivative of **21** providing after deprotection the cardenolide analog **24** in satisfactory yield (*46*).

(6*R*)- Dihydropyranones

A reaction channel to pyranoid enolone esters with only one chiral center left – as compared to the five of the D-glucose starting material – also starts from the readily accessible hydroxyglycal esters, e. g. **8**. Actually, **8** constitutes an ester of the enol form of 1,5-anhydro-D-fructose, in which the more reactive enediol ester group should – under the appropriate conditions – undergo selective saponification, thus liberating the carbonyl function. This cannot be achieved directly though, since even mildly basic conditions result in a range of products originating from β-elimination in the tribenzoate of 1,5-anhydro-D-fructose **26** initially formed, and subsequent formation of the *vic*-diketone and ensuing rearrangements of the benzilic acid type.

8

1,5-Anhydro-D-fructose

88 % NH_2OH

25

MeCHO 90 %

26

NaOAc Me_2CO 92 %

27

A preparatively satisfactory means of selective enol ester cleavage consists in the treatment with hydroxylamine, which not only induces hydroxylaminolysis of the more reactive enol ester group to form the respective hydroxamic acid, but captures the keto group thus liberated in the form of its stable oxime (*29*):

This delightfully simple methodology is generally applicable to hydroxyglycal esters and has provided acylated ketoximes of type **25** in large variety acylated pyranoid ketoximes, featuring such useful properties as high tendency for crystallization, ease of isolation and stability (*29*).

Deoximation of these ketoximes may readily be accomplished by any of the standard procedures, transoximation to acetaldehyde, for example **25** → **26**, being feasible in 90 % yield (*30*). Elimination of benzoic acid is as easily effected, stirring with sodium acetate in acetone at ambient temperature providing the pyranoid enolone ester **27** in crystalline, enantiopure form. The efficiency with which this building block can be elaborated from D-glucose is noteworthy: the six-step reaction sequence involved can be reduced to two hectogram-adaptable one-pot procedures comprising the conversion D-glucose → hydroxyglucal ester (85 %), and the one-pot sequence hydroxylaminolysis → deoximation → elimination (84 %) (*48, 49*).

The versatility of the enolone ester **27** is illustrated by a variety of synthetically useful addition reactions, e. g. with lithium alkyls, Grignard reagents and cyclopentadiene in a Diels-Alder type fashion (*40*, *48*, *49*), but, most notably, it could be utilized for straightforward syntheses of the marine natural products (-)-bissetone (**28**), a metabolite from the Gorgonian soft coral *Briareum polyanthes* (*50*), and (-)-palythazine (**29**), an unusual dipyranopyrazine isolated from the salt water invertebrate *Palythoa tuberculosa* (*51*).

28
Bissetone
(*Briareum polyanthes*)

29
Palythazine
(*Palythoa tuberculosa*)

That their absolute configuration is *S,S* in each case was only established by their acquisition in enantiopure form from building block **27**, its chirality in the single asymmetric center fortuitously being a perfect match.

(*S,S*)-Bissetone (**28**), when traced back to building block **27**, only lacks the 3-carbon branch, i. e. acetone. Indeed, the lithium enolate of acetone proved to be a suitable three-carbon synthon attacking the carbonyl function with a 4:1 preference from the proaxial side (**27** → **30**). The benzoyl group shift directly following the attack elaborates the desired 2-oxopropyl-branched tetrahydropyranone **31**, the dibenzoate of bissetone in fact, from which the parent compound is generated simply by de-*O*-benzoylation (*48*):

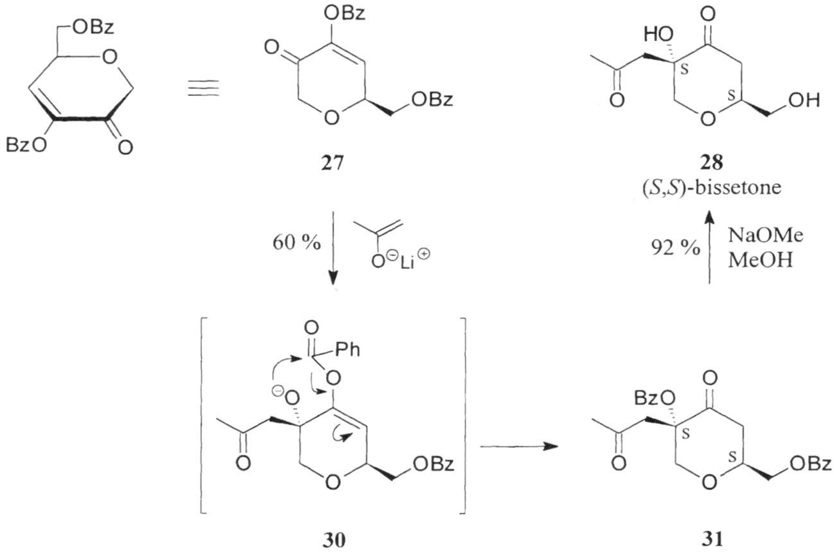

Elaboration of (*S,S*)-palythazine (**29**) from the key building block **27** was similarly effected in a high-yielding reaction sequence: conversion into oxime **32**, liberation of the carbonyl function by debenzoylation (→ **33**), and controlled catalytic hydrogenation to the aminoketone **34**, which dimerizes at pH 9; the concluding step is an air oxidation of the dihydropyrazine initially formed (*52*).

Both synthetic targets, i. e. **28** and **29**, were obtained in crystalline form and their configurational identity with the soft coral-derived products was established on the basis of rotational values, CD and ORD curves, thus unequivocally proving their *S,S*-configuration.

D-Glucose — 6 steps ⟹ **27** — $NH_2OH{\cdot}HCl$, pyr., 96 % → **32**

32 — 84 %, NaOMe/MeOH → **33**

33 — 10% Pd,C/H_2, EtOAc → [**34**]

34 — 69 %, pH 9 (air) → **29**

29
(*S*,*S*)-palythazine

(5*R*)- and (5*S*)-Hydroxyhexanals

With macrolides of the phoracantholide and diplodialide type – the former a defensive secretion from the metasternal gland of the Eucalyptus longicorn beetle *Phoracantha synonyma* (*53*), the latter a metabolite of a plant pathogenic fungus *Diplodia pinea* of high hydroxylase inhibitory activity (*54*) – it may be presumptuous to think of sugars as suitable starting materials.

Diplodialides and Phoracantholides

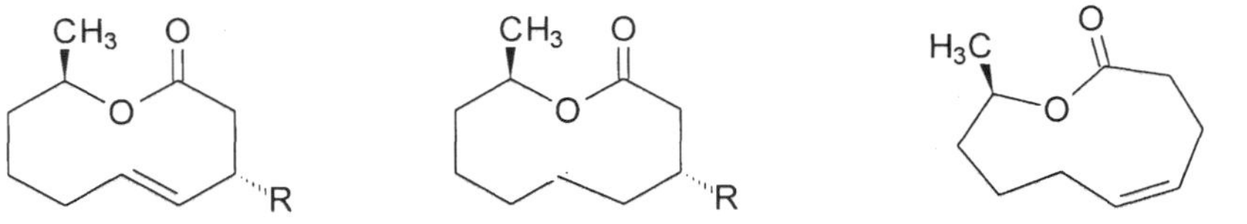

R = H or OH

R = H or OH

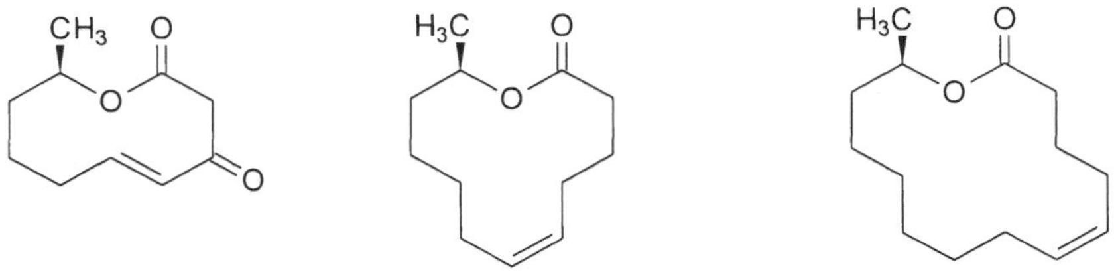

However, there invariably is a single chiral center along the carbon-chains, an (*R*)-hydroxy group tied up as a lactone, and since (5*R*)-hydroxyhexenal **36** or its a cis → trans isomerization product **37** can readily be elaborated from D-glucose in the form of the cyclic acetal **35**, a sugar-based approach to natural products of this type is not only appropriate, but efficient.

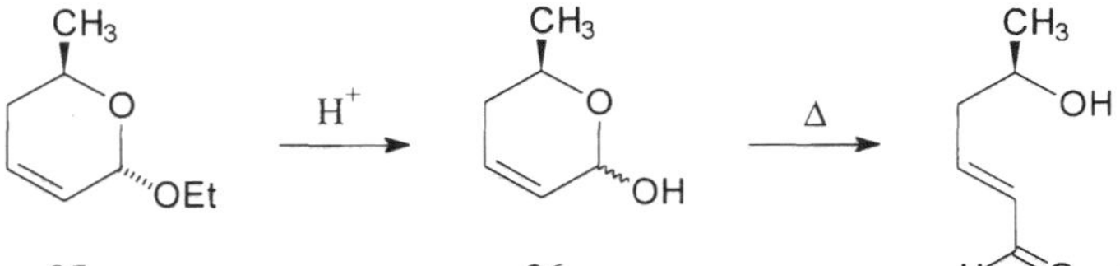

H^+ Δ

35 **36** **3**

These key intermediate is accessible from triacetyl-D-glucal **38** in a large-scale adaptable 5-step sequence in an overall yield of 30 %: conversion of **38** into hexenoside **39** according to a known two-step procedure (*55*), tosylation to the chloro-tosylate **40**, since the chloride ions formed during the reaction in situ displace the activated allylic tosyloxy function, and, finally, the consecutive reductive removal of tosyloxy- and chloro groups **40** → **35** (*56,57*):

The high versatility of **38** as a key intermediate towards macrodiolides is amply demonstrated by its ready hydrolysis to the respective hexenal **40** or, when preceded by hydrogenation, to the 5*R*-hydroxyhexanal **39**, whilst

D-Glucose → (3 steps, 70 %) → **38** → (1. BF_3/EtOH, 2. Et_3N, 70 %) → **39** → (TsCl, 86 %) → **40** → ($LiAlH_4$, 71 %) → **35**

OAc
O
OAc
AcO

OH
O
HO
OEt

OTs
Cl
O
OEt

CH_3
O
OEt

≡

30 % over
7 steps

Approach to Macrodiolides

35

89 % Ni/H_2, H^+ → 41

quant. H^+/Aceton, Δ → 37

81 % BF_3, MCPBA → 42

Diplodialides
Phoracantholides

Carbonolide

Colletodiol

peroxidation yields (*R*)-parasorbic acid **41** (*56, 57*) – all enantiomerically pure 6-carbon building blocks that represent major segments of macrolides: half of diplodialide- and phorocantholide-type pheromones, the C_{11}-C_{16}-segment of carbonolide, or the left portion of colletodiol.

For *C*-extension in the (5*R*)-hydroxyhexanal case it was found (*58*), that the pyranoid lactol form **41** is so stable that reaction with Wittig ylides cannot be effected under standard conditions. Thus, the acylic form had to be elaborated, which was effected by thioketalization, acetylation (**41** → **43**) and desulfurization, the resulting 5-*O*-blocked hexanal **44** then smoothly affording the olefin **45**. Liberation of the hydroxyl function and ensuing lactonization

41 → 43: 1. HS–(CH$_2$)$_3$–HS/BF_3 2. Ac_2O 90 %

43 → 44: $CdCO_3$/MeI 86 %

44 → 45: tBuOK/TH -78 °C 79 %

45 → 46: 1. NaOEt/EtOH 2. Lactonisation 71 %

46 → 47: Pd/H_2 93 %

47
Phoracantholide I
(-35.1°)

46
Phoracantholide J
(-35.9°)

45

according to the Gerlach-modification of the Corey-Nicolaou procedure (*59*) gave phoracantholide J (**46**), subsequent hydrogenation its dihydro derivative, phoracantholide I (**47**) (*57, 58*), the rotations of both being practically identical with those reported (*60*) for the products (*53*) secreted by the Australian eucalyptus longicorn beetle:

Using the same approach, the diplodialides A, B and C, all constituents of the fungi *Diplodia pinea*, have been synthesized in enantiopure form (*58*). This again amply demonstrates the versatility of building blocks of type **41** and **42**, of which the enantiomeric (5*S*)-hydroxy-analogs **51** and **53**, in fact, are as readily accessible from l-rhamnose via di-*O*-acetyl-L-rhamnal **48** (*61*): conversion into the 2,3-unsaturated ethyl glycoside **49**, and standard tosylation to the 4-chloro compound **50**, since the 4-*O*-tosylate of **49** primarily formed undergoes S_N2 displacement by chloride ion under the reaction conditions; the outcome of the subsequent dechlorination depends on the reagent used, $LiAlH_4$ reduction proceeding with conjugate addition of the hydride ion to afford dihydropyrane **51**, whereas exposure to $NiCl_2/NaBH_4$ in aqueous ethanol smoothly elaborates the isomeric **52** (*61*):

Each of the dihydropyranes thus obtained can be converted into their respective enelactones by BF_3-catalyzed peroxidation (*62*), the hexenoside **52** then providing *S*-parasorbic acid **53** (*61*).

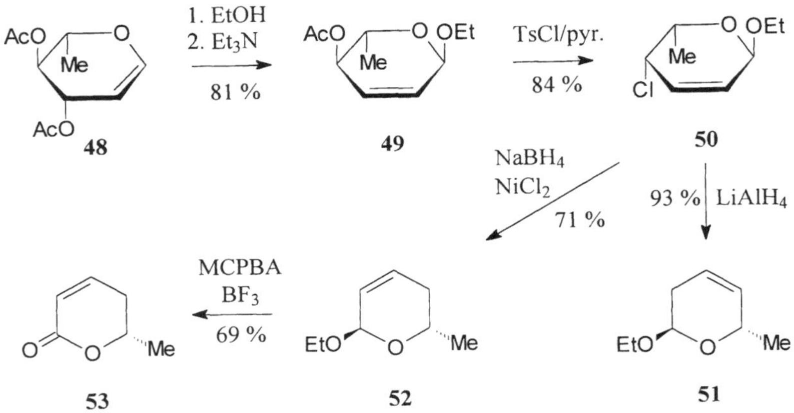

3-C-Methyl-D-allose

Another highly versatile building block derived from diacetone-glucose **54** is the 1,2-acetonide of 3-*C*-methyl-α-D-allose in its furanoid form **57**, which has been utilized as the key compound in a convergent total synthesis of ACRL Toxin I (*63*). Its elaboration from **54** starts with a pyridinium dichromate / acetic anhydride oxidation (*64*), is followed by carbonyl olefination of the respective 3-ulose with methyl (triphenyl)phosphonium bromide and hydrogenation (→ **55** → **56**), and is completed by acid cleavage of the 5,6-isopropylidene group. This four-step process **54** → **57**, upon optimization of reaction conditions and work-up procedure, allows an overall yield of 58 % (*63*), as compared to the 22 % obtained previously (*65*).

Further processing of **57** towards the ketone **60** is readily effected by highly regioselective tosylation of the primary hydroxyl group (*66*), hydride reduction **58** → **59**, and oxidation with pyridinium chlorochromate (PCC) on aluminum oxide to afford **60** in a yield of 70 % over the three steps (*63*). Due to the now practical accessibility of these furanoid building blocks supplementary modifications, that have already been performed, become preparatively relevant, e. g. the conversion of tosylate **58** into the 5,6-epoxide (*66*), *C*-extensions (*63*, *66*), shortening of the carbon chain via periodation of **57** (*63*), and transformation of the respective products into acyclic derivatives by acid hydrolysis of the 1,2-*O*-isopropylidene group (*63*, *66*).

The utilization of the furanoid 3-*C*-methyl-D-allose building blocks **57** and **60** for a convergent total synthesis of ACRL Toxin I in the form of its stable 3-*O*-methyl ether (*63*) involved their conversion into enantiomerically uniform connective segments. The key feature of the retrosynthesis was the expectation

1. PDC
2. $Ph_3PMe/BuLi$
74 %

$Pd,C/H_2$
85 %

84 % H^+

TsCl/pyr.
93 %

PCC/Al_2O_3
93 %

OH, Me, HO, X, O

54 **55** **56**

60 **58** X = OTs → **59** X = H **57**

ACRL Toxin I
61 R = H
62 R = Me

A
(X = PPh_3I,SO_2Ph)

B

C

60

57

that the dihydro-α-pyrone ring could be introduced via a suitable acetoacetic ester derivative. Thus, the molecule was dissected into the segments **A**, **B**, and **C**, the advantage of this segmentation being a convergent reaction sequence, since both, the twofold methyl-branched C_6-chain of synthon **A** and the adipaldehyde building block **B** are derivable – each in the correct absolute configuration – from the 1,2-acetonide of 3-deoxy-3-*C*-methyl-α-D-allofuranose (**57**):

This strategy has been realized as outlined providing the ACRL Toxin I in the form of its 3-*O*-methyl ether **61** (*63*), a variety of contiguous and non-contiguous stereocenters can advantageously be elaborated by relying on the predictable regio- and stereochemistry provided by the carbon framework of furanoid or, sequentially, acyclic sugar-derived building blocks. The preparative outcome of the synthesis is quite satisfactory: 16 % overall yield for the thirteen steps from diacetone glucose to the C_5-C_{10} building block B and 10 % for the preparation of the C_{11}-C_{16}-segment A (as the sulfone) from the same starting material. The overall sequence shows good levels of stereocontrol in installing the six asymmetric centres and two stereodefined double bonds, corresponding to 2.5 steps per stereogenic unit.

Whilst the ACRL Toxin I totally synthesized in this way is the main and most toxic component of the toxins produced by the phytopathogenic fungus *Alternaria citri* (*68*), other synthetic work has concentrated on a minor, essentially non-toxic component, i. e. ACRL Toxin IIIb (**63**) (*69*, *70*) – a task considerably easier as lack of a stereocenter in the pyranoid ring greatly facilitates the joining of the acyclic and pyranoid fragments.

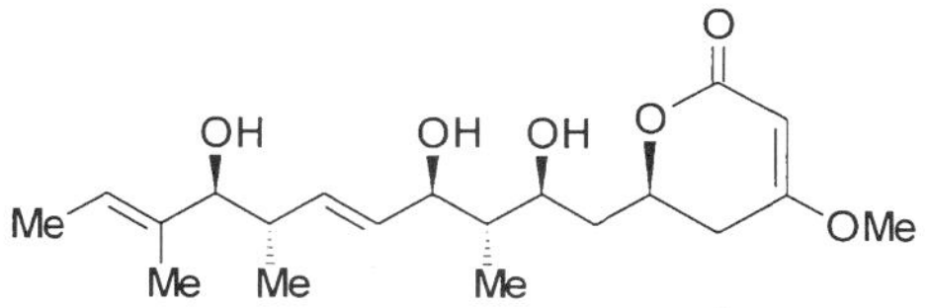

ACRL Toxin I
(**61**)

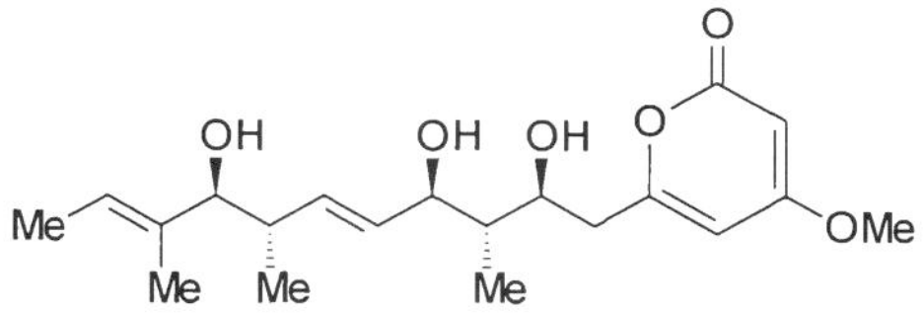

ACRL Toxin III
(**63**)

Modified Cyclodextrins

The use of low molecular weight carbohydrates as starting materials for the generation of versatile enantiopure building blocks may also be extended to the bulk scale-accessible disaccharides such as sucrose (*10*), lactose, maltose and isomaltulose (*71*) as well as to higher oligosaccharides, most notably the cyclodextrins, which are well accessible from starch on a ton scale level (*72*). Their use, in this context, lies less in the elaboration of building blocks to be incorporated into complex target molecules, but in the design of novel flexible host molecules with which to study and eventually understand recognition phenomena at a molecular level.

In the cyclodextrins readily obtainable from starch, the six (α-CD), seven (β-CD) and eight (γ-CD) α(1→4)-linked glucose units are "locked up" in a strait-jacket type belt due to adoption of $^4C_1$ chair conformations of the pyranoid rings and a net of 2-OH ···· OH-3' hydrogen bonds (*73*, *74*). As this structural rigidity even persists on inclusion complex formation, as exemplified by the three represenatives in Fig. 1 (*75* - *78*):

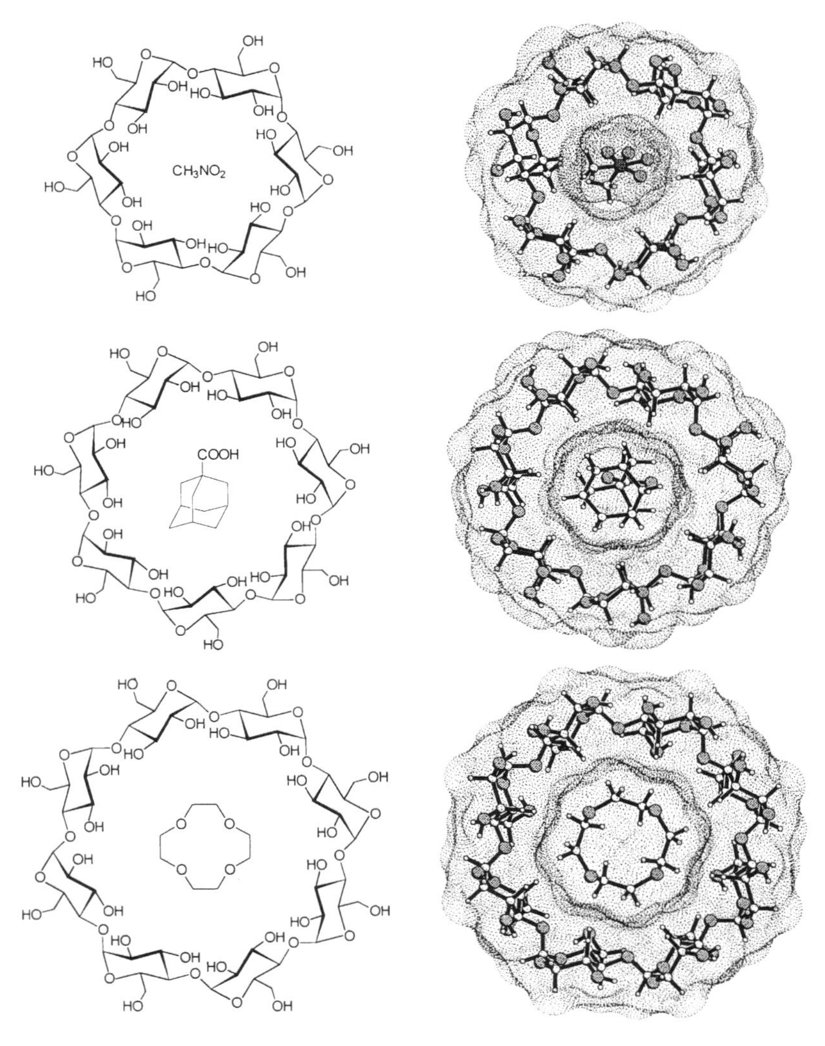

Fig. 1. Topographies of cyclodextrin inclusion complexes: α-CD – nitromethane (*75*) (top), β-CD – adamantane-carboxylate (*76*, *77*) (center), and γ-CD – 12-crown-4 (*76*, *78*). The striking correspondence of hydrophobic surface regions of guest and CD-host at their interfaces may be viewed in color on the Internet (*79*).

Accordingly, incorporation of guests by α-, β-, and γ-CD closely correspondends to Emil Fischer's classic lock-and-key concept for enzyme specificity (*80*), symbolizing the insertion of a lipophilic key into an equally lipophilic cyclodextrin molecular lock. Although overly static, this process has been extensively exploited towards 'artificial enzymes' or 'enzyme models' (*81*), despite of overwhelming evidence that the majority of enzymes act in an induced fit' fashion (*82*), implying the induction of significant conformational changes upon 'docking' of the substrate. Such a distinctly dynamic process obviously is essential for the catalytic groups to assume the required transition state geometry.

By consequence, if low molecular weight cyclodextrins are to be realistic enzyme models, flexibility has to be introduced into their macrocycles so that they can mimic the dynamic induced fit mode of action rather than the stationary lock-and-key approach.

"Installation" of flexibility into the common cyclodextrins – i.e. without altering their α–(1→4)-intersaccharidic link up – implies configurational changes within their glucose moieties. Inversion of one equatorial OH group, however, is not sufficient, as e.g. α-cyclomannin (the axial mannose-2-OH pointing away from the cavitiy (*83*)) and α-cycloallin (the axial 3-OH of the allose units directed into the cavity (*84*), cf. Fig. 3) retain the rigid $^4C_1$ chair conformation in their pyranoid rings, resulting in altered, yet still inflexible topographies. The same holds for per-2,3-anhydro-α-cyclomannin (Fig. 2, top right), in which the pyranose units adopt $^0H_5$ halfchair conformations due to the 2,3-oxirane rings (*85*), yet its cavity is capable to include guests such as ethanol (*85*) or 1-propanol (*86*).

To really embody flexibility into the common cyclodextrins without altering their α-(1→4) interglycosidic link-up requires configurational changes in the glucopyranoid ring at two positions, at C-2 and C-3 in fact, as in the resulting α-cycloaltrin the $^4C_1$ and $^1C_4$ chair conformations of the α-D-altropyranose units are energetically equivalent and, hence, are in equilibration with each other via their $^0S_2$ skew-boat form:

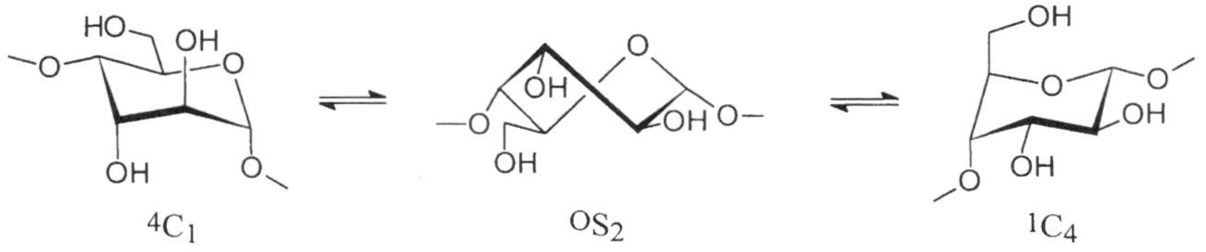

HO OH O OH O — $^4C_1$ ⇌ O OH O OH OH O — $^0S_2$ ⇌ OH O O OH O OH — $^1C_4$

Readily prepared from α-CD in four high-yielding steps (*87*), α-cycloaltrin crystallizes in a disk-shaped topography devoid of an open-ended cavity (Fig. 3, left), since the altrose units are alternately arranged in essentially perfect $^4C_1$ and $^1C_4$ conformations:

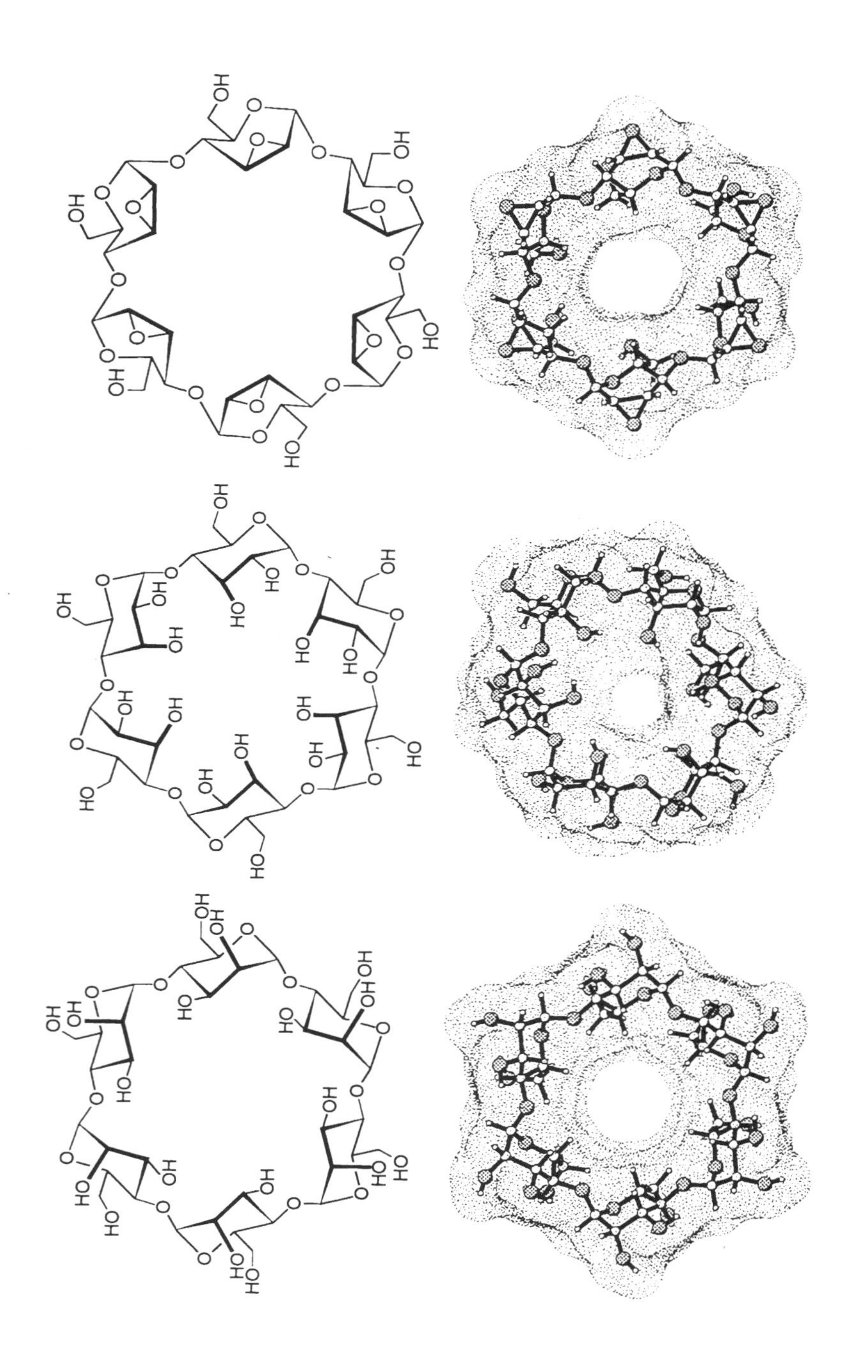

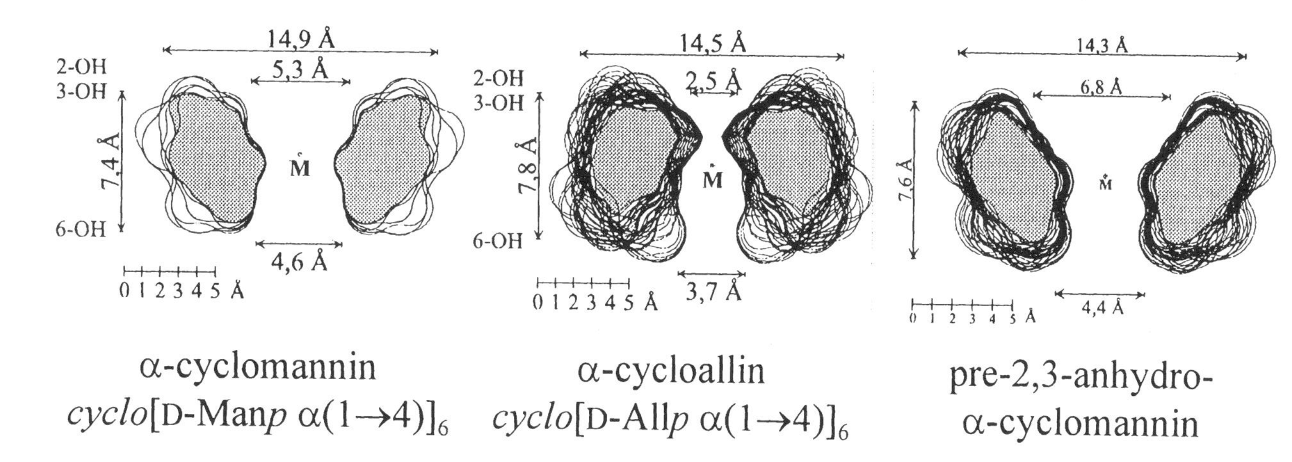

Fig. 2. Topographies of hexameric non-glucose cyclooligosaccharides composed of D-mannose (*83*), D-allose (*84*) and 2,3-anhydro-D-mannose units (*85*) in an α-(1→4)-link-up each.

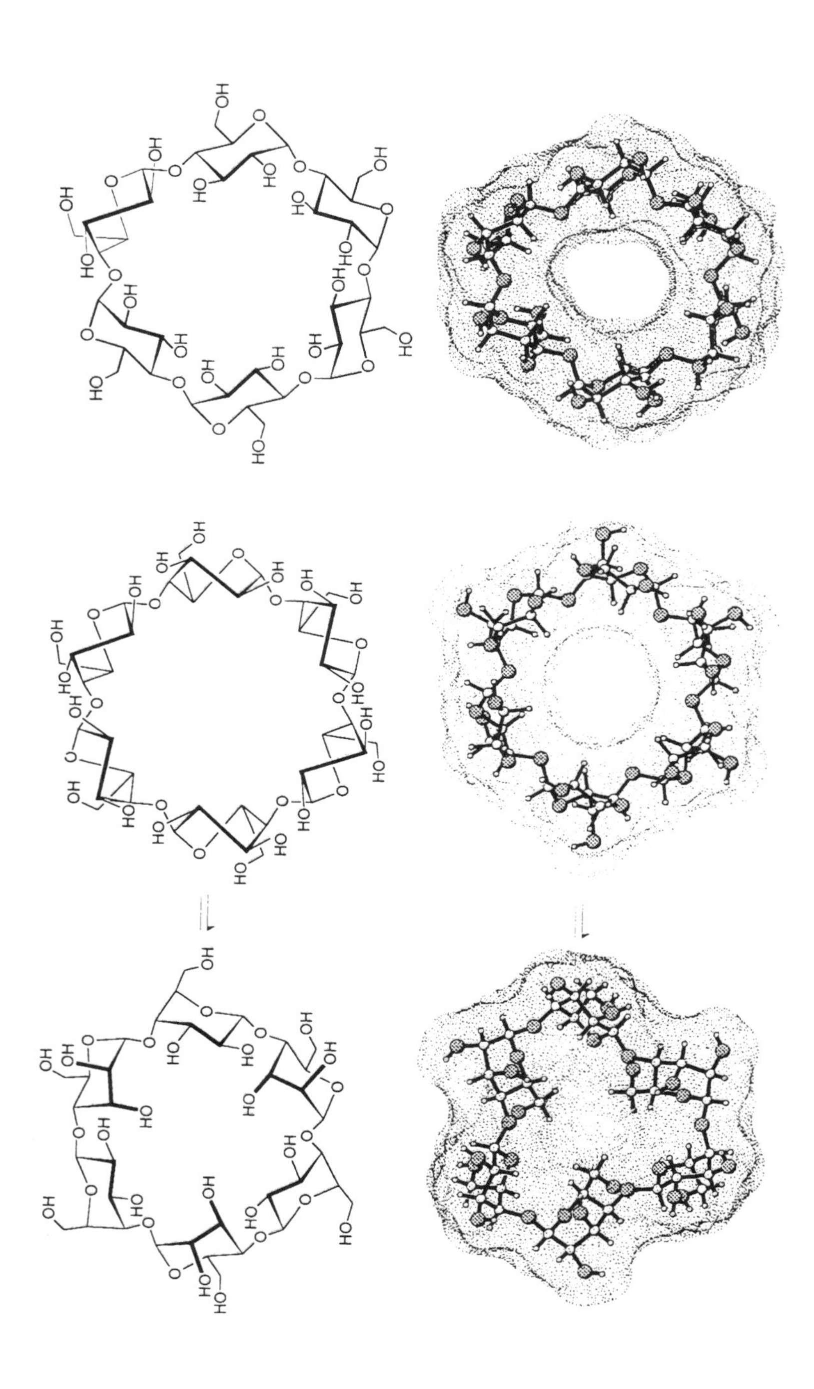

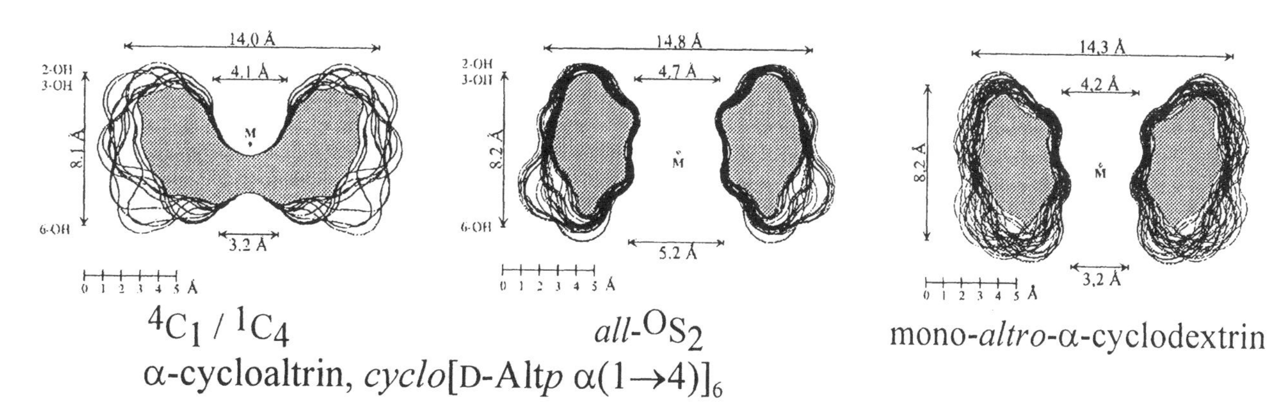

Fig. 3. The two extreme moelcular shapes of α-cycloaltrin (left and center) between which a complex equilibrium is established in solution (*84*). Right: an α-CD analog in which one of the six glucose units is converted to a flexible altrose residue.

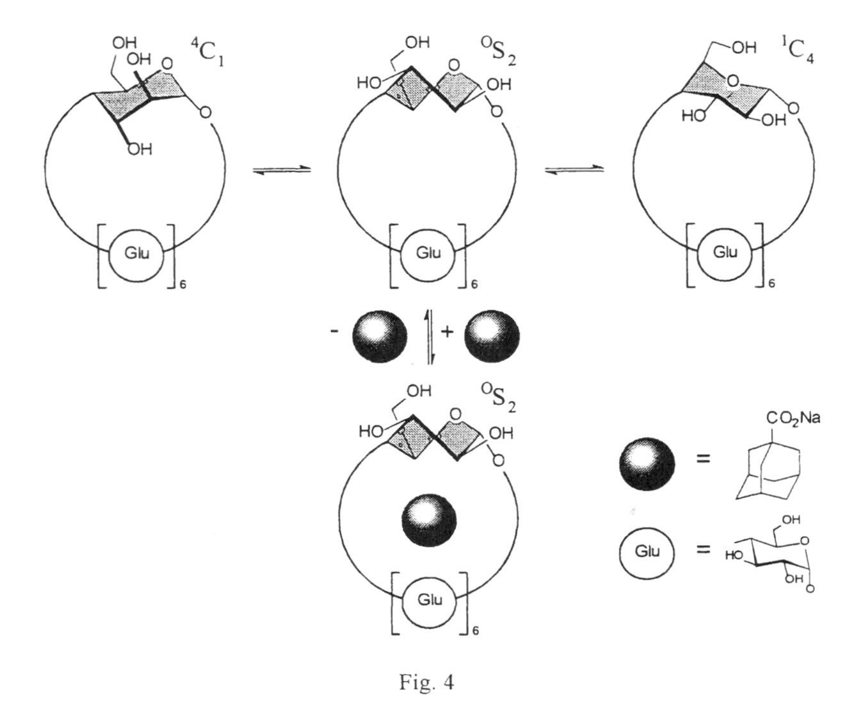

Fig. 4

In aqueous solution, however, temperature-dependent ^1H and ^{13}C NMR studies, togehter with molecular dynamics simulations, reveal the six altropyranose units to be in a complex dynamic equilibrium within the $^4C_1 \rightleftharpoons {}^0S_2 \rightleftharpoons {}^1C_4$ pseudorotational turntable. This gives rise to a large number of macrocyclic conformations, ranging from a disk-shaped molecule with a hydrophobic central indentation (Fig. 3, left) to a torus form of the macrocycle with an equally hydrophobic open-ended cavity, when all pyranoid rings are in the skew-boat $^0S_2$ form (*85*). Accordingly, α-cycloaltrin, and similarly its β- and γ-analogs (*88*), constitute the first thoroughly flexible cyclooligosaccharides with which to realistically probe the induced-fit mode (*82*) of guest-host interactions – a reasonable expectation since mono-*altro*-β-cyclodextrin, featuring one altrose and six glucose residues in an α(1-4)-link-up (Fig. 4, left), can adapt the conformation of its flexible altrose moiety to the adamantane-carboxylate guest (*89*):

Coda

The exemplary *reaction channels* described in this account, leading from simple sugars to versatile enantiopure building blocks, provide a conceptual framework by which the chemistry of low-molecular weight carbohydrates, accessible in bulk quantities, can be moulded towards the practical elaboration of complex natural or otherwise interesting products in enántiomerically homogeneous, optically active form. The synthetic potential, however, inherent in cheap, bulk-scale available carbohydrates is huge and far from being exhausted. Thus, particularly in view of the comparatively few reaction sequences meeting process chemistry demands, there is an urgent necessity to further develop practical, large-scale adaptable *reaction channels* from sugars to versatile building blocks – a task that can successfully be achieved only if the present chemical methodology is utilized to its fullest and the increasingly emerging biotechnological procedures as well. All of this, unambiguously, points towards broad-scale, practicality-oriented basic research to be performed not only in academic institutions, but also in industrial laboratories, most effectively, of course, if both cooperate closely. In short, the challenges for the 21st century, at least in outline form, are clear. The capacity to develop vibrant and inciteful collaborations between academic and industrial institutions is likely to emerge as one of the new frontiers of the utilization of carbohydrates which Nature offers us on an annual basis.

Acknowledgments

My greatest appreciation goes to my scientific partners, my students and postdoctoral associates, who are individually acknowledged in the references. Continuing support from the Deutsche Forschungsgemeinschaft, the Ministry of Agriculture (via its Fachagentur Nachwachsende Rohstoffe), the Fonds der Chemischen Industrie, and the Südzucker AG Mannheim/Ochsenfurt, provided the sustenance for these investigations.

References

1. *Carbohydrates as Organic Raw Materials;* Lichtenthaler, F. W., Ed.; VCH Publ.: Weinheim/New York, **1991**, 367 pp.
2. *Carbohydrates as Organic Raw Materials II*; Descotes, G., Ed.; VCH Publ.: Weinheim/New York, **1993**, 278 pp.
3. *Nachwachsende Rohstoffe: Perspektiven für die Chemie*; Eggersdorfer, M.; Warwel, S.; Wulff, G., Eds.; VCH Publ.: Weinheim/New York, **1993**, 402 pp.
4. *Carbohydrates as Organic Raw Materials III*; van Bekkum, H.; Röper, H.; Voragen, A. G. J., Eds.; VCH Publ.: Weinheim/New York, **1996**, 315 pp.
5. *Nachwachsende Rohstoffe: Perspektiven für die Chemie*; Eierdanz, H., Ed.; VCH Publ.: Weinheim/New York, **1996**, 358 pp.
6. Lichtenthaler, F. W.; Mondel, S. *Pure Appl. Chem.* **1997**, *69*, 1853-1866.

7. Lichtenthaler, F. W., Utility of Ketoses as Organic Raw Materials, *Carbohydr. Res.* **1998**, *313*, 69-89.
8. Hugill, A. in *Sugar and All That. A History of Tate and Lyle*; Introductory Dedicational Metaphor; Gentry Books: London, **1978**.
9. Khan, R. *Adv. Carbohydr. Chem. Biochem.* **1976**, *33*, 235. – Jenner, M. R. in *Developments in Food Carbohydrates, 2*; Lee, C. K., Ed.; Applied Science Publ.: London, **1980**, 91-143.
10. Lichtenthaler, F. W.; Pokinskyj, P.; Immel, S. *Zuckerind. (Berlin)* **1996**, *121*, 174-190.
11. World Sugar Production 2000/2001, *Zuckerind. (Berlin)* **2001**, *126*, 300, 577.
12. Hanessian, S. *Total Synthesis of Natural Products*, Pergamon Press: Oxford, **1983**.
13. Lichtenthaler, F. W. In *Modern Synthetic Methods*; Scheffold, R. Ed.; VCH Publ.: Weinheim/New York, **1992**, 6, 273-376.
14. Bols, M. *Carbohydrate Building Blocks*; Wiley: New York, **1996**, 179 pp.
15. Lichtenthaler, F. W. *Carbohydr. Res.* **1998**, *313*, 69-89.
16. For reviews see: Jurczak, J.; Pikul, S.; Bauer, T. *Tetrahedron* **1986**, *42*, 447-488. – Altenbach, H. J. *Nachr. Chem. Tech. Lab.* **1988**, *36*, 33-38.
17. Fischer, E. *Ber. Dtsch. Chem. Ges.* **1894**, *27*, 673-679. – Wolfrom, M. L.; Thompson, A. *Methods Carbohydr. Chem.* **1963**, *2*, 427-430.
18. Fischer, E., *Ber. Dtsch. Chem. Ges.* **1895**, *28*, 1145-1167. – Schmidt, O. T. *Methods Carbohydr. Chem.* **1963**, *2*, 318-325.
19. Fischer, E. *Ber. Dtsch. Chem. Ges.* **1893**, *26*, 2400-2412. – Bollenback, G. N. *Methods Carbohydr. Chem.* **1963**, *2*, 327-328.
20. Fischer, E.; Zach, K. *Ber. Dtsch. Chem. Ges.* **1914**, *47*, 196-210. – Roth, W.; Pigman, W. *Methods Carbohydr. Chem.* **1963**, *2*, 405-408.
21. Maurer, K. *Ber. Dtsch. Chem. Ges.* **1929**, *62*, 332-338. – Maurer, K.; Petsch, W. *Ber. Dtsch. Chem. Ges.* **1933**, *66*, 995-2000. – Ferrier, R. J. *Methods Carbohydr. Chem.* **1972**, *6*, 370-311.
22. Koenigs, W.; Knorr, E. *Ber. Dtsch. Chem. Ges.* **1901**, *34*, 957-981.
23. Fischer, E.; Armstrong, E. F. *Ber. Dtsch. Chem. Ges.* **1901**, *34*, 2885-2900.
24. Redemann, C. E.; Niemann, C. *Org. Syntheses, Coll. Vol. III*, **1955**, 11-14. – Lemieux, R. U. *Methods Carbohydr. Chem.* **1963**, *2*, 221-222.
25. Lichtenthaler, F. W.; Cuny, E.; Weprek, S. *Angew. Chem.* **1983**, *95*, 906-907; *Angew. Chem. Int. Ed. Engl.* **1983**, *22*, 891-892.
26. Lichtenthaler, F. W.; Kläres, U.; Lergenmüller, M.; Schwidetzky, S. *Synthesis* **1992**, 179-184.
27. Lichtenthaler, F. W.; Kraska, U. *Carbohydr. Res.* **1977**, *58*, 363-377.
28. Lichtenthaler, F. W. in *Modern Synthetic Methods*, Scheffold, R. Ed.; VCH Publ.: Weinheim/New York, **1992**, *6*, 360-361.
29. Lichtenthaler; F. W. Jarglis, P. *Tetrahedron Lett.* **1980**, *21*, 1425-1428.
30. Lichtenthaler, F. W., El Ashry, E. S. H., Göckel, V. H. *Tetrahedron Lett.* **1980**, *21*, 1429-1432.

31. Lichtenthaler, F. W.; Jarglis, P. *Angew. Chem.* **1982**, *94*, 643; *Angew. Chem. Int. Ed. Engl.* **1982**, *21*, 625. *Angew. Chem. Suppl.* **1982**, 1449-1459. – Lichtenthaler, F. W., Jarglis, P.; Hempe, W. *Liebigs Ann. Chem.* **1983**, 1959-1972.
32. Lichtenthaler, F. W.; Kaji, E.; Weprek, S. *J. Org. Chem.* **1985**, *50*, 3503-3515.
33. For an early review, see: Kaji, E.; Lichtenthaler, F. W. *Trends Glycoscience Glycotechnol.* **1993**, *5*, 121-142.
34. Lichtenthaler, F. W.; Schneider-Adams, T. *J. Org. Chem.* **1994**, *59*, 6728-6734. – Lichtenthaler, F. W.; Kläres, U.; Szurmai, Z.; Werner, B. *Carbohydr. Res.* **1997**, *305*, 293-303. – Lichtenthaler, F. W.; Metz, T. W. *Tetrahedron Lett.* **1997**, *38*, 5477-5480.
35. Lichtenthaler, F. W.; Schwidetzky, S.; Nakamura, K. *Tetrahedron Lett.* **1990**, *31*, 71-74.
36. Binch, H. M.; Griffin, A. M; Schwidetzky, S.; Ramsay, M. V. J., Gallagher, T.; Lichtenthaler, F. W. *J. Chem. Soc., Chem. Commun.* **1995**, 967-968.
37. Cuny; E., Lichtenthaler, F. W. **1997/98**, unpublished results.
38. Lichtenthaler, F. W.; Kraska, U.; Ogawa, S. *Tetrahedron Lett.* **1978**, 1323-1326.
39. Lichtenthaler, F. W.; Nishiyama, S.; Köhler, P.; Lindner, H. J. *Carbohydr. Res.* **1985**, *136*, 13-26.
40. Dauben, W. G.; Kowalczyk, B. A.; Lichtenthaler, F. W. *J. Org. Chem.* **1990**, *55*, 2391-2398.
41. Rosenbrook, Jr., W. *J. Antibiotics* **1979**, *32*, S 211-S 227, and literature cited there.
42. Büschweiler, F.; Stöckel, K.; Reichstein, T. *Helv. Chim. Acta* **1969**, *52*, 2276-2303.
43. Lichtenthaler, F. W.; Löhe, A.; Cuny, E. *Liebigs Ann. Chem.* **1983**, 1973-1985.
44. Hanessian, S.; Roy, R. J. J. *Am. Chem. Soc.* **1979**, *101*, 5839-5841.
45. White, D. R.; Birkenmeyer, R. D.; Thomas, R. C., Mizsack, S. A.; Wiley, V. H. 44. *Tetrahedron Lett.* **1979**, 2737-2740.
46. Lichtenthaler, F. W.; Cuny, E.; Sakanaka, O., unpublished results 1997-1999.
47. Lindner, H. J.; Lichtenthaler, F. W., to be published.
48. Brehm, M.; Dauben, W. G., Köhler, P.; Lichtenthaler, F. W. *Angew. Chem.* **1987**, *99*, 1318-1319; *Angew. Chem. Int. Ed. Engl.* **1987**, *26*, 1271-1273.
49. Lichtenthaler, F. W., in *Modern Synthetic Methods*; R. Scheffold, Ed.; VCH Publ.: Weinheim/New York, **1992**, *6*, 308-310 and359.
50. Cardellina II, J. H.; Hendrickson, R. L.; Manfred, K. P.; Strobel, S. A.; Clardy, J. *Tetrahedron Lett.* **1987**, *28*, 727-730.
51. Uemura, D.; Toya, I.; Watanabe, I.; Hirata, Y. *Chem. Lett.* **1979**, 1481-1484.

52. Jarglis, P.; Lichtenthaler, F. W. *Angew. Chem.* **1982**, *94*, 140-141; *Angew. Chem. Int. Ed. Engl.* **1982**, *21*, 141-142; *Angew. Chem. Suppl.* **1982**, 175-184.
53. Moore, B. P.; Brown, W. V. *Aust. J. Chem.* **1976**, *29*, 1365-1374.
54. Ishida, T.; Wada, K. *J. Chem. Soc., Chem. Commun.* **1975**, 205-206.
55. Ferrier, R. J.; Prasad, N. *J. Chem. Soc. C* **1969**, 570-575. – Fraser-Reid, B.; McLean, A.; Usherwood, E. W.; Yunker, M. *Can. J. Chem.* **1970**, *48*, 2877-2884.
56. Lichtenthaler, F. W., in *New Aspects in Organic Chemistry* (Z. Yoshida, T. Shiba, Y. Ohshiro, Eds.), VCH Publ.: Weinheim/New York, **1989**, 372 ff.
57. Klingler, F. D. *Doctoral Dissertation*, Technische Universität Darmstadt, **1985**.
58. Neff, K. H. *Doctoral Dissertation*, Technische Universität Darmstadt, **1988**.
59. Gerlach, H.; Künzler, P.; Oertle, K. *Helv. Chim. Acta* **1978**, *61*, 1226-1231. – Skenvi, A. B.; Gerlach, H. *ibid.* **1980**, *65*, 2426-2431.
60. Kitahara, T.; Koseki, K.; Mori, K. *Agric. Biol. Chem.* **1983**, *47*, 389-393.
61. Lichtenthaler, F. W.; Klingler, F. D.; Jarglis, P. *Carbohydr. Res.* **1984**, *132*, C1-C5.
62. Lichtenthaler, F. W.; Rönninger, S.; Jarglis, P. *Liebigs Ann. Chem.* **1989**, 1153-1161.
63. Lichtenthaler, F. W.; Dinges, J.; Fukuda, Y. *Angew. Chem.* **1991**, *103*, 1385-1389; *Angew. Chem. Int. Ed. Engl.* **1991**, *30*, 1339-1343. – An extensive discussion of the strategy for and the execution of the total synthesis of (-)-ACRL toxin I was given by Hale, K.J., in *Organic Synthesis with Carbohydrates*; Boons, G-J.; Hale, K.J., Eds.; Sheffield Acad. Press: Sheffield, UK, **2000**, pp 270-279.
64. Anderson, F.; Samuelson, B. *Carbohydr. Res.* **1984**, *129*, C4-C3.
65. Rosenthal, A.; Sprinzl, M. *Can. J. Chem.* **1969**, *47*, 3941-3946.
66. Kinoshita, M.; Mariyama, S. *Bull. Chem. Soc. Jpn.* **1975**, *48*, 2081-2083.
67. Dinges, J.; Lichtenthaler, F. W.; Lindner, H. J. *Z. Kristallogr.* **1994**, *209*, 512-513.
68. Gardner, J. M.; Kono, Y.; Tatum, J. H. Suzuki, Y.; Takeuchi, S. *Phytochemistry* **1985**, *24*, 2861-2867. – Kono, Y.; Gardner, J. M. Suzuki, Y; Takeuchi, S. *Phytochemistry* **1985**, *24*, 2869-2874. – Kono, Y.; Gardner, J. M.; Kobayashi, S.; Sakurai, T. *Phytochemistry* **1986**, *25*, 69-72.
69. Mulzer, J.; Dupré, S.; Buschmann, J.; Luger, P. *Angew. Chem.* **1993**, *105*, 1538-1540; *Angew. Chem. Int. Ed. Engl.* **1993**, *32*, 1452-1454.
70. Munchhof, M. J.; Heathcock, C. H. *J. Org. Chem.* **1994**, *59*, 7566-7567.
71. Lichtenthaler, F. W.; Immel, S.; Martin, D.; Müller, V. *Starch/Stärke* **1992**, *44*, 445-456.
72. Szejtli, J. *Chem. Rev.* **1998**, *98*, 1743-1753.
73. Harata, K. *Chem. Rev.* **1998**, *98*, 1803-1827. – Lipkowitz, K. B. *ibid.* **1998**, *98*, 1829-1873.
74. Immel, S.; Lichtenthaler, F. W. *Liebigs Ann. Chem.* **1996**, 27-37.

75. Nakagawa, T.; Immel, S.; Lichtenthaler, F. W.; Lindner, H. J. *Carbohydr. Res.* **2000**, *324*, 141-146.
76. Lichtenthaler, F. W.; Immel, S. *Starch/Stärke* **1996**, *48*, 145-154.
77. Hamilton, J. B.; Sabersan, M. N. *Acta Cryst.* **1982**, *B32*, 3063-3069.
78. Kamitori, S.; Hirotsu, K.; Higuchi, T. *J. Chem. Soc., Chem. Commun.* **1986**, 690-691.
79. All 3D-structures can be viewed at "http://caramel.oc.chemie.tu-darmstadt.de/immel/3Dstructures.html". – The color-coded lipophilicity patterns (MLP's) are available at "http://caramel.oc.chemie.tu-darmstadt.de/immel/molcad/gallery.html.
80. Fischer, E. *Ber. Dtsch. Chem. Ges.* **1894**, *27*, 2985-2993. – Lichtenthaler, F. W. *Angew. Chem.* **1994**, *106*, 2456-2467; *Angew. Chem. Int. Ed. Engl.* **1994**, *33*, 2364-2374.
81. Breslow, R.; Dong, S. D. *Chem. Rev.* **1998**, *98*, 1997-2011.
82. Koshland Jr., D. E. *Angew. Chem.* **1994**, *106*, 2468-2472; *Angew. Chem. Int. Ed. Engl.* **1994**, *33*, 2375-2378.
83. Lichtenthaler, F. W.; Immel, S. *Tetrahedron: Asymmetry* **1994**, *5*, 2045-2060. – *J. Incl. Phenom. Mol. Recogn. Chem.* **1996**, *25*, 3-16.
84. Immel, S.; Fujita, K.; Lichtenthaler, F. W. *Chem. Eur. J.* **1999**, *5*, 3185-3192.
85. Immel, S.; Fujita, K.; Lindner, H. J.; Nogami, Y.; Lichtenthaler, F. W. *Chem. Eur. J.* **2000**, *6*, 2327-2333.
86. Immel, S.; Lichtenthaler, F. W.; Lindner, H. J.; Fujita, K.; Fukudome, M.; Nogami, Y. *Tetrahedron: Asymmetry* **2000**, *11*, 27-36.
87. Nogami, Y.; Nasu, K.; Koga, T.; Ohta, K.; Fujita, K.; Immel, S.; Lindner, H. J.; Schmitt, G. E.; Lichtenthaler, F. W. *Angew. Chem.* **1997**, *109*, 1987-1981; *Angew. Chem. Int. Ed. Engl.* **1997**, *36*, 1899-1902.
88. β-Cycloaltrin: Fujita, K.; Shimada, H.; Ohta, K.; Nogami, Y.; Nasu, K. *Angew. Chem.* **1995**, *107*, 1783-1784; *Angew. Chem. Int. Ed. Engl.* **1995**, *34*, 1621-1622. – γ-Cycloaltrin: Nogami, Y.; Fujita, K.; Ohta, K.; Nasu, K.; Shimada, H.; Shinohara, C.; Koga, T. *J. Inclusion Phenom. Mol. Recogn. Chem.* **1996**, *25*, 53-56.
89. Fujita, K.; Chen, W.-H.; Yuan, D.-Q.; Nogami, Y.; Koga, T.; Fujioka, T.; Mihashi, K.; Immel, S.; Lichtenthaler, F. W. *Tetrahedron: Asymmetry* **1999**, *10*, 1689-1696.

Chapter 5

General Three Carbon Chiral Synthons from Carbohydrates: Chiral Pool and Chiral Auxiliary Approaches

Rawle I. Hollingsworth* and Guijun Wang

Department of Chemistry and Biochemistry, Michigan State University, East Lansing, MI 48824
***Corresponding author: rih@cem.msu.edu**

The use of carbohydrates to gain access to chiral 3-carbon synthons for use in applications including pharmaceuticals, agrichemicals, biomaterials and certain advanced materials is discussed. Three basic approaches are described. The first one is based on the direct scission of carbohydrates to yield 3-carbon synthons directly, the second on 1 carbon degradation reactions on optically pure 3-hydroxy-γ-butyrolactones or their derivatives and the third approach is based on the use of carbohydrates as chiral auxiliaries to induce chirality into pro-chiral substrates. Because of the general availability of optically pure 3-hydroxy-γ-butyrolactones on commercial scale and the ease of the transformations, the second approach has several practical advantages and represents a significant step forward our drive to develop a general carbohydrate-based chemistry platform.

Introduction

Carbohydrates have the highest density of functionality and chiral centers of all naturally occurring molecules. Some carbohydrates such as lactose, sucrose, glucose and maltose are readily available in very pure crystalline forms on commodity scale. Some polymers such as starch, chitin and cellulose are produced as primary and secondary agricultural products. Carbohydrates are also obtained from renewable resources. Because of these facts, carbohydrates have long held the ultimate promise as raw materials for the development of a general chemistry where a high density of chemical functionality and high optical purity are required. A promise that can in no way be matched by fossil fuel chemistries.

Despite all of this promise, there are some very real challenges in the development of a carbohydrate-based chemistry. The main ones arecomplexity and redundancy. The high frequency of hydroxyl groups and the high density with which they are packed into the molecular skeleton makes it very difficult to develop orthogonal derivitization methods for mono and disaccharides and virtually impossible for anything larger than a trisaccharide. Because of this, the chemistry of carbohydrate polymers such as starch and cellulose has been one in which physical attributes such as viscosity, gelling properties, swelling properties, stickiness and other decidedly material properties are manipulated. This is usually accomplished by changing polarity by techniques such as methylation or changing charge by technicques such as carboxymethylation or sulfation. There has not been much success in confronting the complexity of these molecules to generate useful substructures with functionality and optical purity to be of use in areas like pharmaceuticals, agrichemicals and advanced materials until comparatively recently. The industrial chemistry of starch has been reviewed. (*1*)

Several years ago, we introduced a method for the preparation of optically pure 3,4-dihydroxybutyric acids and their γ-lactones based on the alkaline oxidation of 4-linked hexoses (*2-4*).

The method (scheme 1) was targeted to the utilization of readily available carbohydrate sources such as lactose, maltose starch and maltodextrins. It was applicable largely to the preparation of the (S)-acid **1** and the lactone **2** since naturally occurring hexoses have almost exclusively the D-configuration and the C5 carbon becomes the chiral center in the dihydroxy acid. More recently, direct access to the (R)-isomers **4** and **5** was made possible by the development of a similar oxidation method using 5-linked pentoses as the starting compounds iinstead (*5*) (scheme 2). The availability of these 4-carbon compounds with both stereochemistries has formed the basis for a very rich chemistry that has been extended to the preparation of optically pure 3-carbon compounds. This will be the main focus of the following discussion.

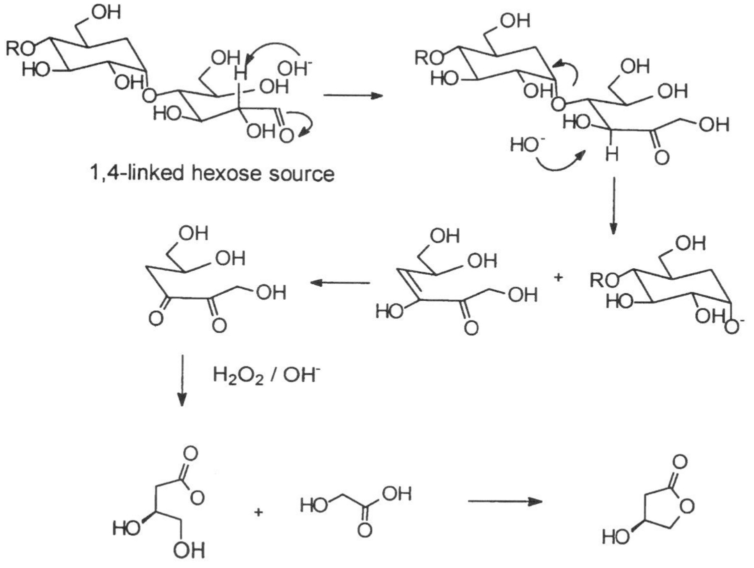

1,4-linked hexose source

H_2O_2 / OH^-

1

2

Scheme 1

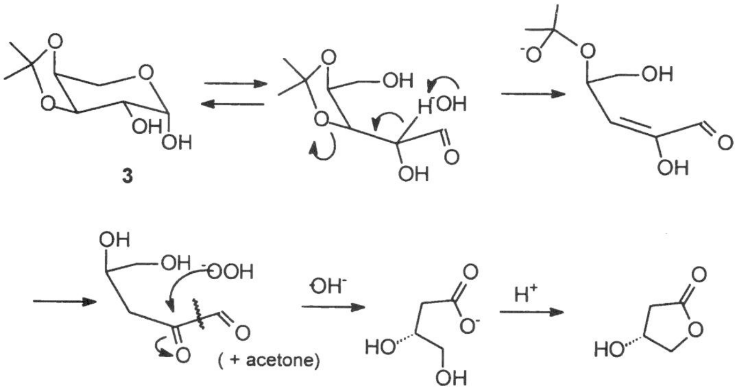

(+ acetone)

OH^-

H^+

3

4

5

Scheme 2

Optically pure 3-carbon compounds from carbohydrates

There are several methods for the preparation of optically pure 3-carbon compounds from carbohydrates. These fall into 3 major categories namely the cleavage of carbohydrate chains to form the desired compounds directly, the transformation of a (larger) carbohydrate-derived fragment to a 3-carbon one and the use of carbohydrates as a chiral auxiliary to induce chirality into a pro-chiral 3-carbon fragment. We will review the first approach and discuss our efforts in the other two areas more fully.

Oxidative scission of carbohydrates directly to 3-carbon synthons

Isopropylidene and other alkylidene glyceraldehydes

The oxidative scission of an protected saccharide is perhaps the oldest of the approaches for preparing chiral 3-carbon synthons from carbohydrates. The most used method is the oxidation of 1,2-5,6-di-O-isopropylidene-D-mannitol **6** to yield 2,3-O-isopropylidene-D-glyceraldehyde **7** using either periodate or lead tetra-acetate (scheme 3) (*6*). The cyclohexylidene group is often used in place of the isopropylidene. This reaction usually proceeds with satisfactory yield on small scale but has little or no commercial value for several reasons. These include the high molecular weights of the reagents, the tendency for over-oxidation to the carboxylic acids and the toxicity of lead.

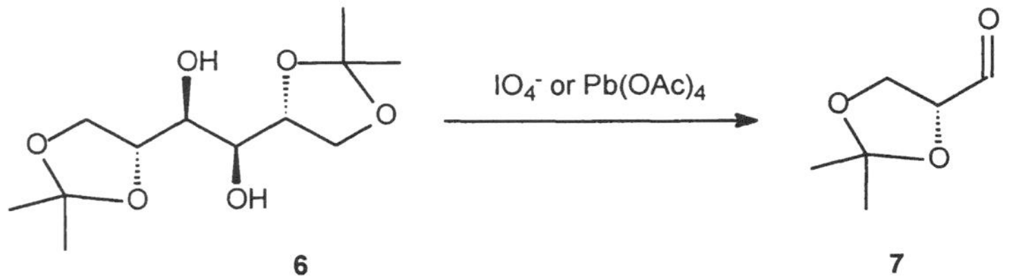

IO_4^- or $Pb(OAc)_4$

6 **7**

Scheme 3

Access to the isomeric (L)-protected glyceraldehydes by this approach is limited because of the general unavailability of L-mannitol. One way of circumventing this problem is to employ a 2-step degradation of 5,6-O-isopropylidene-L-ascorbic acid **8** (scheme 4) (*7*).

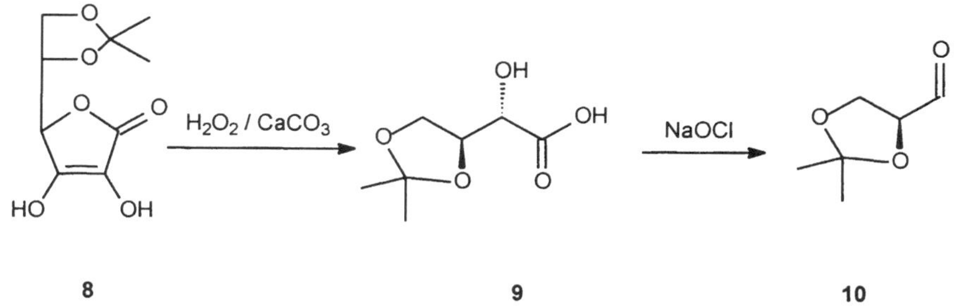

8 9 10

Scheme 4

This affords the L-acetal **10** via hypochlorite treatment of the intermediate 3,4-acetal of L-erythrose **9**. The protected O-alkylidene glyceraldehydes are extremely flexible and can be converted into a plethora of other chiral synthons with a wide selection of uses using very standard chemical transformations (scheme 5). One major limitation for its use is the ease of racemization of the chiral center because of the ease of enolization of the aldehyde group. This is especially true even in the mildest of basic conditions.

Isopropylidene and other alkylidene glyceric acids

A more recent and potentially more useful method (*8*) for preparing optically active 3-carbon synthons is by the ruthenium catalysed oxidation of 1,2-5,6-di-

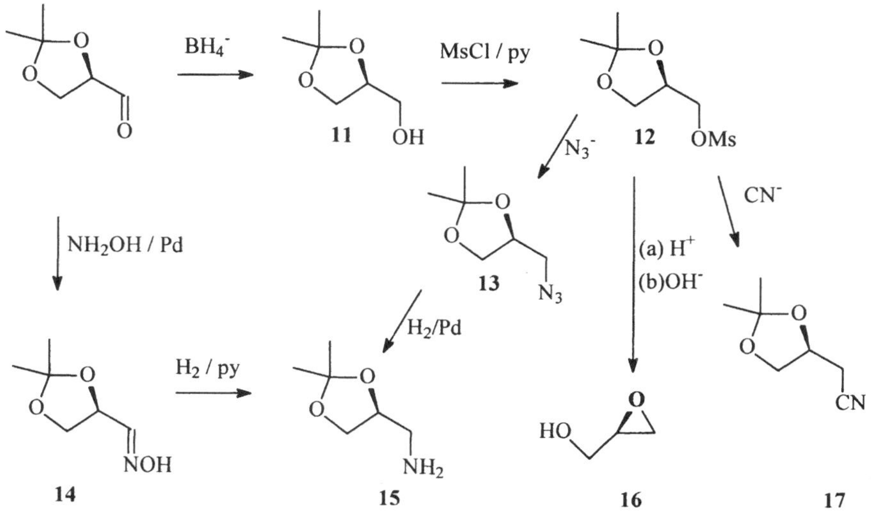

Scheme 5

O-isopropylidene-D-mannitol (scheme 6) to produce isopropylidene D-glyceric acid **18**.

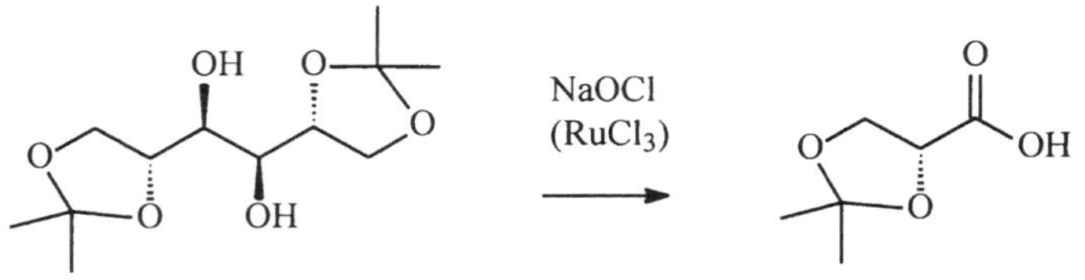

NaOCl
($RuCl_3$)

18

Scheme 6

This has several advantages over the other methods mentioned above chief among which is the much lower cost of reagents and the fact that the use of heavy metals is limited to catalytic amounts. In a similar approach, hypochlorite was used in the presence of perruthenate catalysts to oxidize acetals of ascorbic and iso-ascorbic acid to optically pure acetals of glyceric acid (*9*).

Chiral 3-carbon synthons from optically pure 3,4-dihydroxybutyric acids.

The general availability of optically pure 3,4-dihydroxybutyric acids has opened up several new routes to chiral 3-carbon fragments with a wealth of functionality. The (S)- and (R)-enantiomers of 3-amino-1,2-dihydroxypropane (**15** and **19** respectively) are among the most useful chiral 3-carbon synthons.

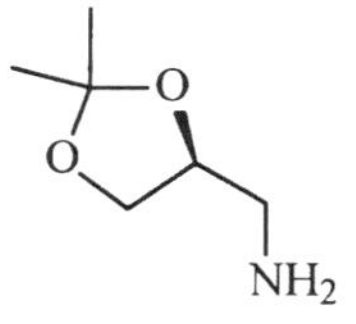

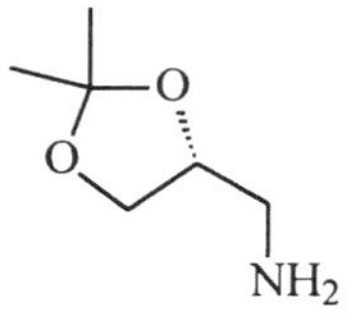

15 **19**

A very facile method of synthesizing them is a Hofmann degradation of the corresponding isopropylidene acetal of 3,4-dihydroxybutyramide (*10*). The amide **20** is obtained from the lactone **2** in quantitative yield by treatment with ammonia. The protection of the dihydroxy function with the isopropylidene group is necessary because participation of the 4-hydroxyl group during the rearrangement will simply yield the starting lactone. The preparation of the (S)-enantiomer **15** by this method is illustrated in scheme 7.

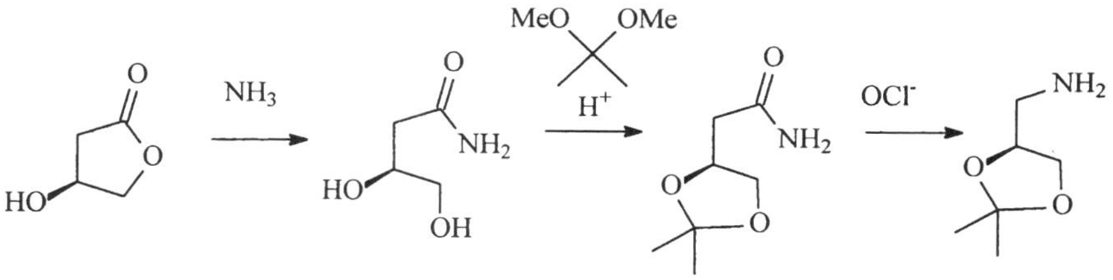

20 21 15

Scheme 7

The acetal can be further transformed into the corresponding bromo-diol and epoxide **22** and **23** respectively (scheme 8).

Another method for producing a chiral 3-carbon fragment, this time directly as a protected 5-hydroxymethyl-3-oxazolidin-2-one, is illustrated in scheme 9 (*11*). In this case, the amide **20** is converted to the 4-trityl ether **24**. This undergoes very facile Hofmann rearrangement to give the 5-trityloxymethyl-3-oxazolidin-2-one **26** via the intermediate isocyanate **25**. The oxazolidinone **26** is a protected version of 3-amino-1,2-dihydroxypropane.

3-carbon chiral synthons using glucose as a chiral auxiliary

The use of chiral molecules as auxiliaries is an old established procedure. In this approach, chirality is induced in a pro-chiral molecule by a combination of steric and bonding effects. The use of carbohydrates as chiral auxiliaries has met with only modest success. The enantioselective cyclopropanation of 2-alkenyl glycosides is worthy of special note although in this case the product is not a 3-carbon intermediate (*12*). Other attempts such as epoxidations (*13*), dihydroxylations (*14*) and brominations (*15*) gave poor results. The primary problems in the use of carbohydrates as chiral auxiliaries are devising a way of attaching the pro-chiral substrate to the auxiliary in a way in which it can be easily removed. The second problem is building in sufficiently severe steric constraints and favorable electrostatic and bonding interactions. The redundancy of hydroxyl groups also makes attachment of the substrate selectively to one position problematic. We have sought to overcome these problems using various strategies.

In one approach (*16*) to the use of glucose as a chiral auxiliary for the preparation of optically active 3-carbon synthons, we devised a strategy where the substrate was attached to the auxiliary initially at one point and then subsequently at a second point during the reaction. This is illustrated in scheme 10 where the preparation of both enantiomers of 1,2-dihydroxypropane (**29** and

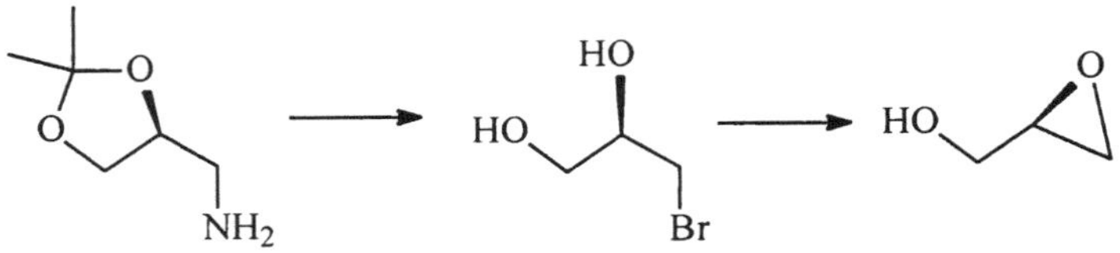

22

23

Scheme 8

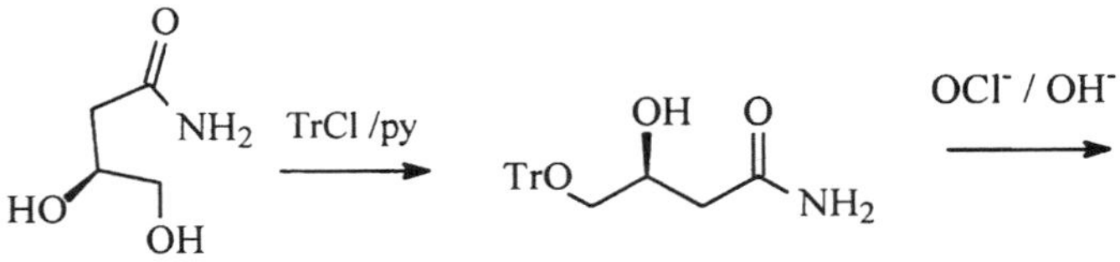

20

24

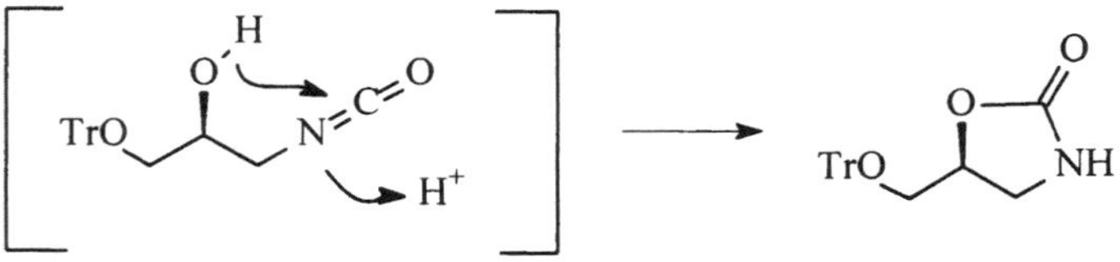

25

26

Scheme 9

31) from allyl alcohol by an oximercuration/de-mercuration sequence is described.

The first point of attachment was through the anomeric position via a glycosidic linkage to form a glycoside with the α- or β- configuration (**27** or**30** respectively). The double bond of the substrate is activated to attack by the 2-hydroxyl group to form a *cis*-fused bicyclic system in the case of the α-glycoside and a *trans* fused system in the case of the β-glycoside. The intermediate mercurated species is demercurated by borohydride reduction. An alternative demercuration strategy is to use bromide instead to give a protected 1-bromo-2,3-dihydroxypropane. The product was recovered by oxidative degradation of the auxiliary. Unlike other chiral auxiliary strategies where both enantiomers of the auxiliary should be available if both enantiomers of the target molecule are to be prepared, only one enantiomer of glucose is required in this method. The enantio-chemistry is decided by which anomer (α– or β-) is used. The driving force that determines the enantiochemical outcome of this reaction is the demanding steric requirements of the fused ring system**28**. The methyl group of the protected propane diol must occupy an equatorial orientation to avoid the extreme steric clash with the axial hydrogen atom at the 3-position. Because of this, very high diastereoselectivities that translated into very high enantioselectivities were obtained.

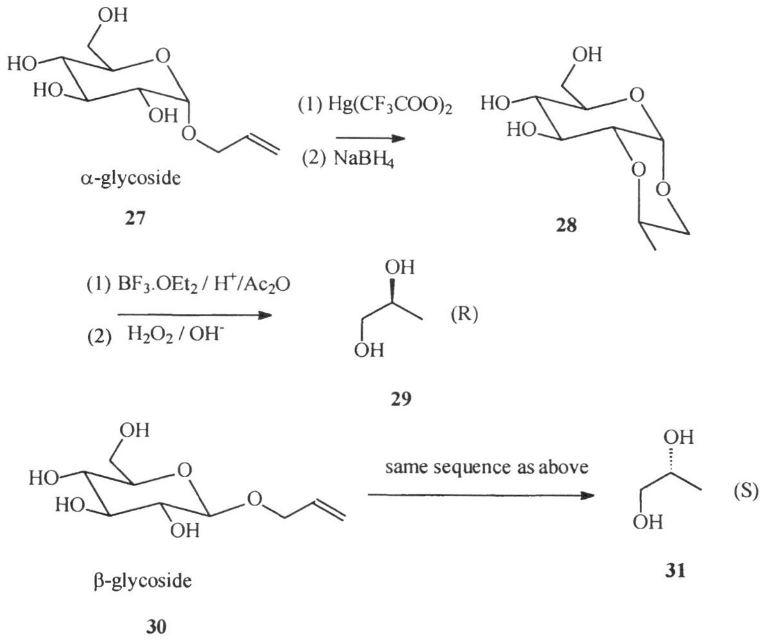

OH HO HO O OH O

α-glycoside

27

(1) $Hg(CF_3COO)_2$

(2) $NaBH_4$

OH HO HO O O O

28

(1) $BF_3.OEt_2$ / H^+/Ac_2O

(2) H_2O_2 / OH^-

OH OH (R)

29

OH HO HO O O OH

β-glycoside

30

same sequence as above

OH OH (S)

31

Scheme 10

One difficulty we encountered in the approach described above was liberation of the 3-carbon fragment from the auxiliary. The glycosidic linkage resisted acid hydrolysis and a rather severe acetolysis followed by basic peroxide oxidation to degrade the carbohydrate moiety had to be performed. A second approach to the use of glucose as an auxiliary is illustrated in scheme 11.

In this approach (17), the prochiral fragment was attached to the auxiliary at only one point making the acid-catalysed release after transformation a simple procedure. Chirality was induced by the borohydride reduction of a carbonyl group on the pro-chiral fragment in the asymmetric environment created by complexation of calcium ions between the C1 and C2 oxygens of the hexose. The diastereomeric purity was good (~70%) but separation of the diastereomers was more problematic.

Applications of chiral 3-carbon synthons

The uses of chiral 3-carbon synthons are many and have been discussed in several reviews and other sources (*18-23*). Molecules that have been prepared from isopropylidene glyceraldehydes include various trialkoxynitrobutanes such as **35** and **36** (*24*), Corey lactone variants **37** and **38** (*25,26*), nucleoside analogs **39-42** (*27)*, verrucarinolactone isomers **43-46** (*28*), the enantiomers of roccellaric acid **47** (29), rubrenolide **48** (*30*) and (+)-laurencin **49** (*31*).

Other compound classes in the synthesis of which chiral 3-carbon synthons were used include sphingosine chains (*32*), 3-amino-2-azetidinones (33), β,γ-unsaturated-α-amino acids (*34*), fluorinated macrocyclic bis(indolyl) maleimides[35], fluorocyclopropyl alcohols (*36*) , 1-O-phosphocholine-2-O-acyl-octadecanes and 1-*O*-phosphocholine-2-N-acyl-octadecane (*37*) diacyl glycerols (*38-42*) and analogs of fragments of leukotriene-B4 (*43*).

Recently oxazolidinones have risen to much prominence as antibacterials for use against various drug-resistant strains (*44-46*) and also for the treatment of behaviour (*47,48*) and neurological (*49,50*) disorders. Optically active isopropylidene glyceraldehyde prepared from both mannitol and ascorbic acids have also been used in the synthesis of 5-hydroxymethyloxazolidinones by two routes, one involving the oxime of the aldehyde and another involving isopropylidene glycerol made by reduction of the aldehyde group (*51*). The method described in scheme 9 gives easy access to this class of compounds. Structures **50-61** are examples of compounds of high pharmacological value that contain chiral 3-carbon fragments.

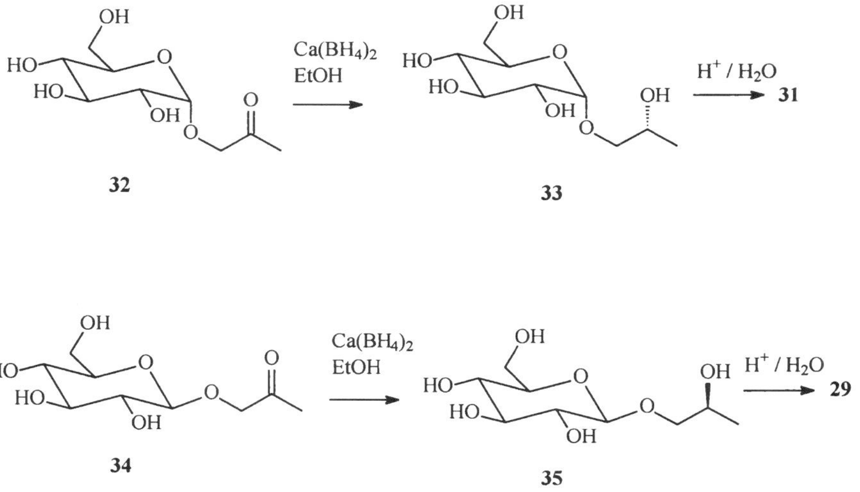

$Ca(BH_4)_2$, EtOH: **32** → **33**; H^+ / H_2O: **33** → **31**

$Ca(BH_4)_2$, EtOH: **34** → **35**; H^+ / H_2O: **35** → **29**

Scheme 11

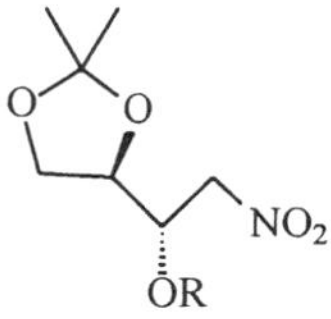

35 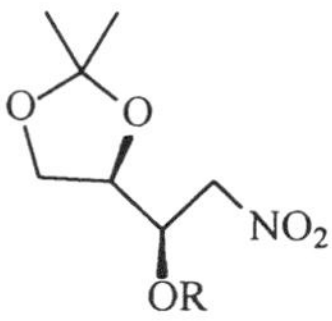 36 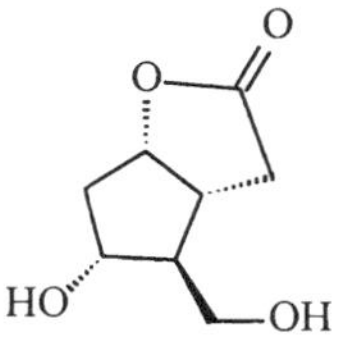 37

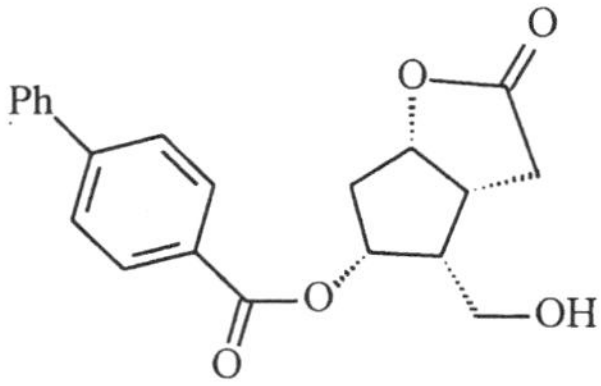

38 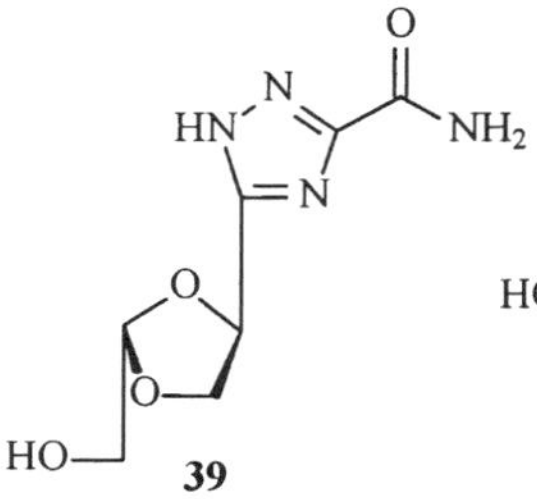 39 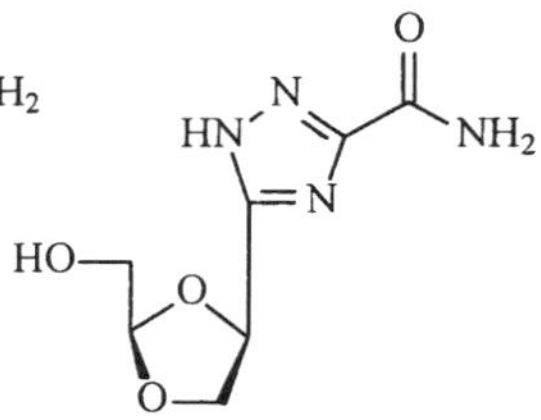 40

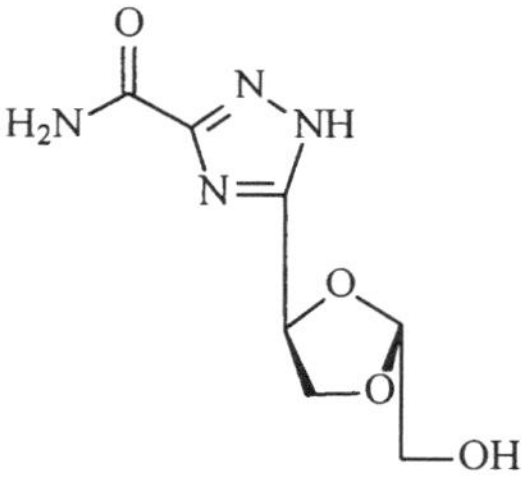

41 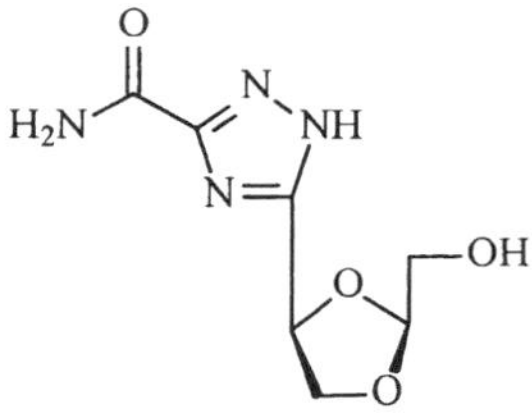 42

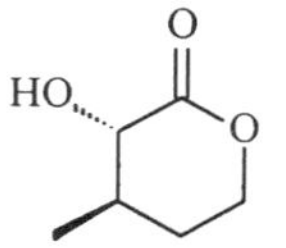

43 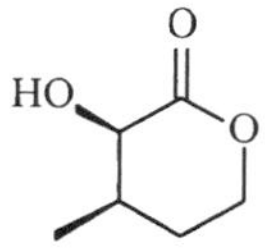 44 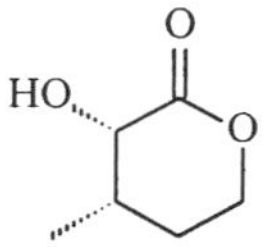 45 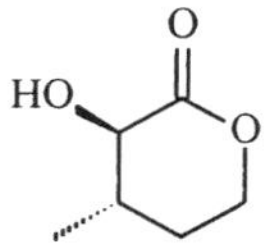 46

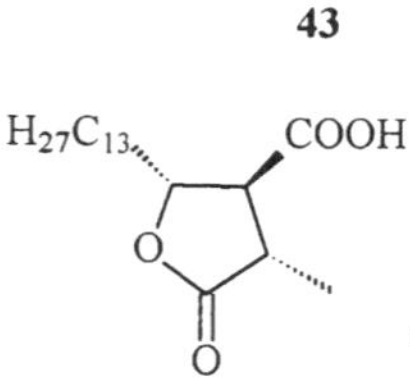

47 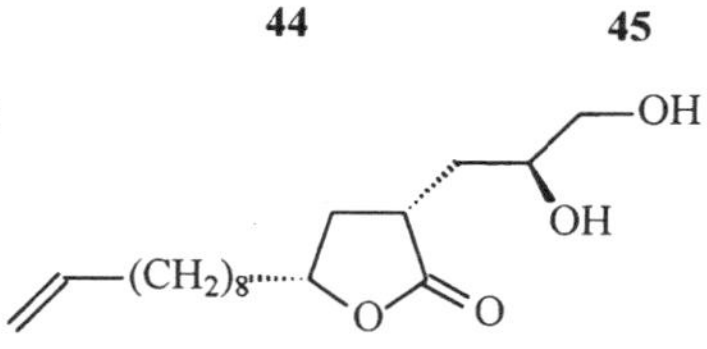 48 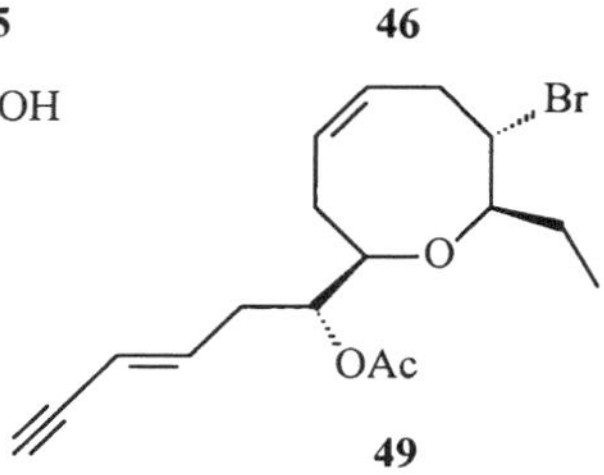 49

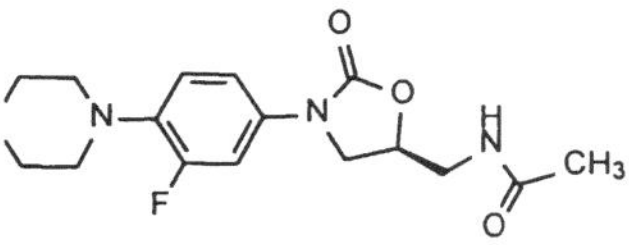

50 Linezolid

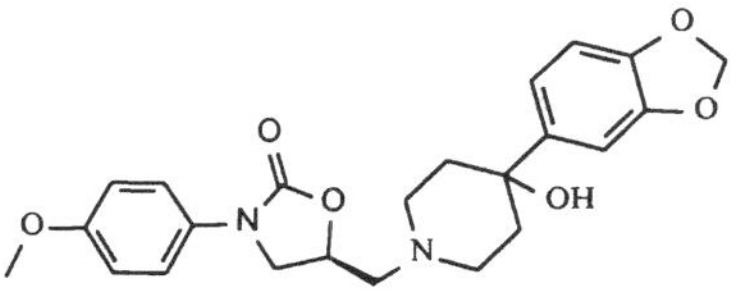

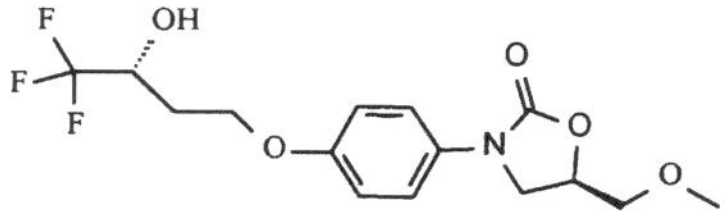

51 Befloxatone

52 Panamasine

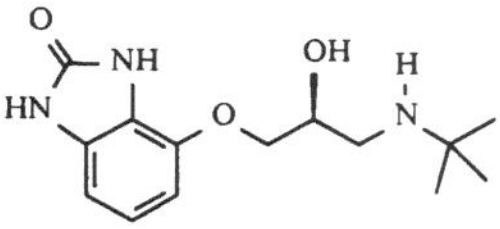

53 beta-3 Adrenergic receptor antagonist for obesity, diabetes

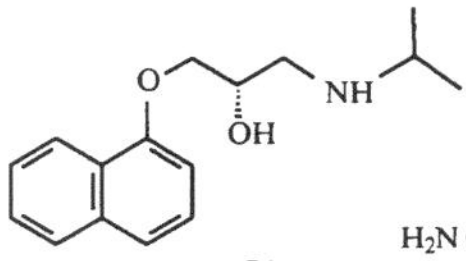

54 *(S)*-Propanol

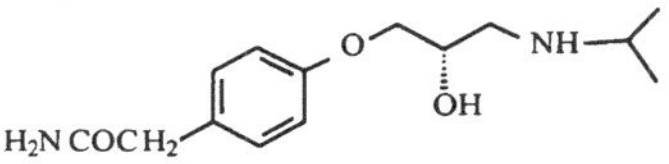

55 *(S)*-Atenolol

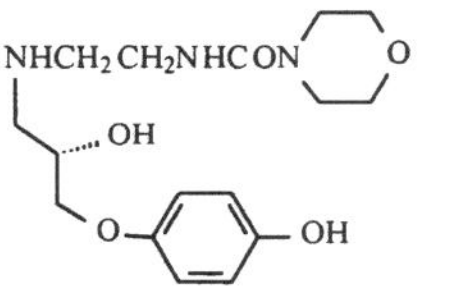

56 Xamoterol, beta-1-Adrenoceptor partial agonist

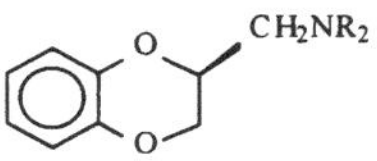

57 alpha-Adrenergic antagonist

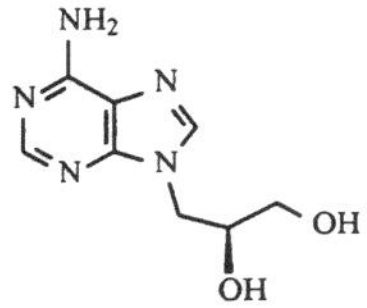

58 Antiviral agent

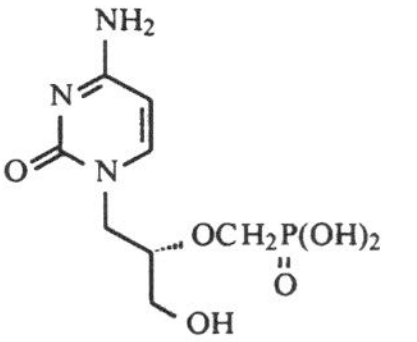

59 Antiviral agent Cidofovir

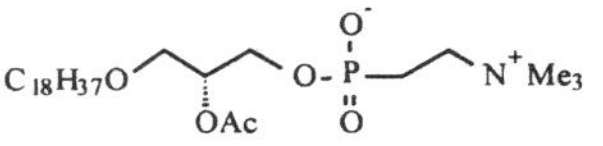

60 Platelet aggregation factor

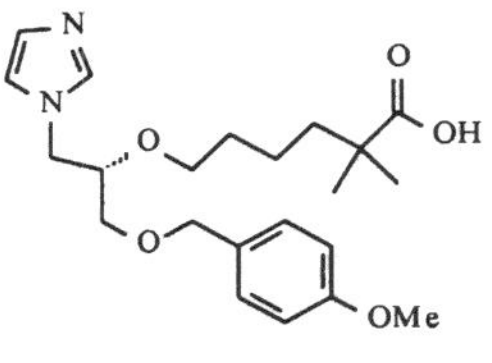

61 Thromboxane synthase inhibitor

Structure **50** is the oxazolidinone antibacterial agent Linezolid, **51** is the oxazolidinone Befloxatone used for treating behaviour disorders, Panasamine **52** (also an oxazolidinone) is used for treating neurological disorders. Compounds **53-56** are β-blockers, **57** is an α-adrenergic antagonist, **58** and **59** are antiviral agents, **60** is a platelet aggregation factor and **61** is a thromboxane synthase inhibitor. These are readily obtainable from the various intermediates derived from optically pure 3-hydroxy-γ-butyrolactones indicated in schemes 8-11.

The most striking feature of structures **50–61** is their sheer diversity and richness and the breadth of medical indications to which they are applicable. The chiral 3-carbon fragment is arguably the most common substructure that can be obtained from a general purpose synthon. Because of this, there has been a steady introduction of methods for preparing chiral 3-carbon fragments into the literature. Hence (R)-glycidol can be prepared from a racemic mixture of glycidyl butyrates by treatment with porcine phospholipases to liberate only this enantiomers (*52*). There are also the Sharpless method for preparing chiral glycidols (*53*) and the Jacobsen chemistry using chiral manganese catalyst for the synthesis of chiral amino-alcohols and epoxides (*54,55*). These are all important additions to the arsenal of synthetic methodologies that chemists have at their disposal.

Conclusion

The main approach we have taken to the application of carbohydrates in chiral chemistry is one of developing as general a chemistry as possible from a few readily-obtainable intermediates. Because we are interested in designing a commercially relevant "green chemistry" strategy, optical purity, yields, cost, environmental impact, flexibility, scalability, directness and generality are all important. In this approach, the stereocenter in the C3 synthons is derived from the 5-position of D-glucose or the 4-position of L-arabinose. Because of this, the optical purities are extremely high (> 99.6 %). Typical yields that have obtained in the conversion of the optically active hydroxy lactones to the 3-carbon intermediates are greater than 85%. Low cost is also a key feature. The carbohydrate raw materials are cheap and available and there are no exotic catalysts. There are also no costly reduction steps. The environmental impact is low because no heavy metals or halogenated solvents are used. The primary carbohydrate raw materials are a renewable resource. The chemistry is very flexible. It is relatively easy to incorporate functionalities such as nitrilo, bromo, chloro, iodo, amino, alkylamino, carboxy and other functionalities into the chiral 3 carbon base fragments we describe here. Scalability is also an important feature of the chemistry. The conversion of carbohydrate raw materials to the intermediate lactone is currently being practiced on the multi ton level. The availability of both isomers contributes to the generality and the direct access to the more highly functionalized chiral 3-carbon synthons such as the amino and halo derivatives is also an important and valuable feature of the approach. In general, ways of generating a palette of chiral compounds without investing in a new specific chiral technology for each are extremely valuable. Capturing the rich structural functionality of carbohydrates is an excellent strategy for doing this

References

1. Hollingsworth, R.I.; *Biotechnology Annual Review* **1996**, *2*, 281-291
2. Hollingsworth, R.I.; *Process for the preparation of 3,4-dihydroxybutanoic acid and salts thereof* U.S. patent 5,292, 939 (March 1994)
3. Hollingsworth, R.I.; *Process for the preparation of 3,4-dihydroxybutanoic acid and salts thereof.* U.S. patent 5,319,110 (June 1994)
4. Hollingsworth, R.I.; *Process for the preparation of 3,4-dihydroxybutanoic acid and salts thereof.* U.S. patents 5,374,773 (December 1994)
5. Hollingsworth, R.I.; *J. Org. Chem.* **1999**, *64*, 7633-7634.
6. Baer, E.; Fisher, H.O.L.; *J. Am. Chem. Soc.* **1948**, *70*, 609
7. Jung, M.E.; Shaw, T.J.; *J. Am. Chem. Soc.* **1980**, *102*, 6304.
8. Emons, C.H.H.; Kuster, B.F.M.; Vekemans, J.A.J.M.; Sheldon, R.A.; *Tetrahedron Asymm.* **1991**, *2*, 359.
9. Carlsen, P.H.J.; Misund, K.; Roe, J.; *Acta. Chem. Scand.* **1995**, *49*, 297.
10. Wang, G.; Hollingsworth, R.I.; *J. Org. Chem.* **1999**, *64*, 1036.
11. Wang, G.; Hollingsworth, R.I.; *Tetrahedron Asymm.* **2000**, *11*, 4429-4432.
12. Charette, A.B.; Cote, B.; Marcoux, J.F.; *J. Am. Chem. Soc.* **1991**, *113*, 8166-8167.
13. Charette, A.B.; Cote, B.; *Tetrahedron Asymm.* **1993**, *4*, 2283-2286.
14. Gurjar, M.K.; Mainkar, A.S. *Tetrahedron Asymm.* **1992**, *3*, 21-24.
15. Bellucci, G.; Chiappe, C.; D'Andreas, F.; *Tetrahedron Asymm.* **1995**, *6*, 221-230.
16. Huang, G.; Hollingsworth, R.I.; *Tetrahedron Lett.* **1999**, *40*, 581-584.
17. Huang, G.; **2000** Ph.D. Thesis, Michigan State University, East Lansing MI, USA.
18. Hanessian, S. *Total Synthesis of Natural Products: The chiron approach;* Pergamon Press, Oxford, **1986.**
19. Hollingsworth, R.I.; Wang, G.J.; *Chim. Oggi* **2000**, *18* (9), 40-42.
20. Hollingsworth, R.I.; Wang, G.; *Chem. Revs.* **2000**, *100*, 4267-4282.
21. Kasai, N.; Suzuki, T.; Furukawa, Y. *J. Mol. Cat. B.* **1998**, *4*, 237-252.
22. Jurczak, J,; Pikul, S.; Tomasz, B. *Tetrahedron* **1986**, *42*, 447.
23. Mulzer, J. *Org. Synth. Highlights* **1991**, 243.
24. Cheng, Q.; Oritani, T.; Hassner, A. *Syn. Comm.* **2000**, *30*, 293.

25. (a) Bindra, J.S.; Bindra, R. *Prostaglandin Synthesis*, Academic Press, New York, **1977**; (b) Collins, P.W. *J. Med. Chem.* **1986**, *29*, 437.
26. Mulzer, J.; Kermanchahi, A. K.; Buschmann, J.; Luger, P. *Liebigs. Ann. Chem.* **1994**, 531.
27. Qu, F. C.; Hong, J. H.; Du, J. F.; Newton, M. G.; Chu, C. K. *Tetrahedron*, **1999**, *55*, 9073.
28. Mulzer, J.; Salimi, N. *Liebigs. Ann. Chem.* **1986**, 1172.
29. Mulzer, J.; Salimi, N.; Hartl, H. *Tetrahedron Asymm.* **1993**, *4*, 457
30. Saito, T.; Thijs, L.; Ettema, G. J.; Zwaneburg, B. *Tetrahedron Lett.* **1993**, *34*, 3589.
31. Mujica, M. T.; Afonso, M. M.; Galindo, A.; Palenzuela, J. A. *Synlett.* **1996**, 983.
32. Mulzer, J.; Brand, C. *Tetrahedron,* **1986**, *42*, 5961.
33. Vandersteen, F. H.; Kleijn, H.; Britovsek, G. J. P.; Jastrzebski, J. T. B. H.; Vanloten, G. *J. Org. Chem.* **1992**, *57*, 3906.
34. Mulzer, J.; Funk, G. *Synthesis,* **1995**, 101.
35. Goekjian, P. G.; Wu, G. Z.; Chen. S.; Zhou, L. X.; Jirousek, M. R.; Gillig, J. R.; Ballas, L. M.; Dixon, J. T. *J. Org. Chem.* **1999**, *64*, 4238.
36. Morikawa, T.; Sasaki, H.; Mori, K.; Shiro, M.; Taguchi, T. *Chem. Pharm. Bull.* **1992**, *40*, 3189.
37. Massing, U.; Eibl, H. *Chem. Phys. Lipids*, **1994**, *69*, 105.
38. Herbert, N.; Beck, A.; Lennox, R.B.; Just, G. *J. Org. Chem.* **1992**, *57,* 1777.
39. Martin, S. F.; Josey, J.A.; Wong, V.L.; Dean, D.W. *J. Org. Chem.* **1994**, *59*, 4805.
40. Martin, S.F.; Josey, A. *Tetrahedron Lett.* **1988**, *29*, 3631.
41. Srisiri, W.; Lee, Y.S.; O'Brien, D.F. *Tetrahedron Lett.* **1995**, *36*, 8945.
42. Srisiri, W.; Lamparski, H.G.; O'Brien D.F. *J. Org. Chem.* **1996**, *61,* 5911.
43. Wang, Y. F.; Li, J. C.; Wu, Y. L. *Acta Chimica Sinica* **1993**, *51*, 409.
44. Bowersock, T. L.; Salmon, S. A.; Portis, E. S.; Prescott, J. F.; Robison, D. A.; Ford, C.W.; Watts, J. L. *Antimicrob. agents and chemotherp.* **2000,** *44*, 1367-1369.
45. Diekema, D. J.; Jones, R. N. *Drugs* , **2000**, *59,* 7-16.
46. Genin, M. J.; Allwine, A.; Anderson, D. J.; Barbachyn, M. R.; Emmert, D. E.; Garmon, S. A.; Graber, D. R.; Grega, K. C.; Hester, J. B.; Hutchinson, D. K.; Morris, J.; Reischer, R. J.; Ford, C. W.; Zurenko, G. E.; Hamel, J. C.; Schaadt, R. D.; Stapert, D.; Yagi, B. H. *J. Med. Chem.* **2000,** *43*, 953-970.

47. Wouters, J.; Moureau, F.; Evrard, G.; Koenig, J.J.; Jegham, S.; George, P., Durant, F.A. *Bioorganic and Medicinal Chemistry* **1999**, 7, 1683-1693.
48. Curet, O.; Damoiseau-Ovens, G.; Sauvage, C.; Sontag, N.; Avenet, P.; Depoortere, H.; Caille D.; Bergis, O.; Scatton, B. *J. Affect. Disorders.* **1998**, 51, 287-303.
49. Frieboes, R.M.; Murck, H.; Antonijevic, I.; Kraus, T.; Hinze-Selch, D.; Pollmacher, T.; Steiger, A. *Psychopharmacology* **1999**, 141, 107-110.
50. Huber, M.T.; Gotthardt, U.; Schreiber, W.; Krieg, J.C. *Pharmacopsychiatry* **1999**, 32, 68-72.
51. Danielmeier, K.; Steckhan, E.; *Tetrahedron Asymm.* **1995,** *6*, 1181
52. Ladner, W.E.; Whitesides G. M. *J. Amer. Chem. Soc.* **1984**, *106*, 7250.
53. Katsuki, T.; Sharpless, K. B.; *J. Amer. Chem. Soc.* **1980**, *102*, 5974.
54. Jacobsen, E.N.; *Acc. Chem. Res.* **2000**, *33*, 421.
55. Wei, E.N.; Loebach, J.L.; Wilson, S.R.; Jacobsen, E.N.; *J. Amer. Chem. Soc.* **1990**, *112,* 2801

Chapter 6

1-Thio-1,2-*O*-Isopropylidene Acetals: Annulating Synthons for Highly Hydroxylated Systems

David R. Mootoo, Xuhong Cheng, Noshena Khan, Darrin Dabideen, Govindaraj Kumaran, and Liang Bao

Department of Chemistry, Hunter College and Graduate Center, CUNY, 695 Park Avenue, New York, NY 10021

Analogues of complex saccharides, in which one or other of the acetal oxygens is replaced by a methylene residue or other heteroatoms, have received attention as biochemical probes and as potential therapeutic agents. The use of 1-thio-1,2-isopropylidene acetals (TIA's) as annulating synthons for highly hydroxylated systems is illustrated by the synthesis of β-C-, β-aza-C- and β-carba- galacto disaccharides.

Introduction

The implication of carbohydrate mechanisms in a number of health disorders has led to interest in glycomimetic structures as biochemical probes of carbohydrate-receptor interaction, and as potential therapeutic agents (*1-3*). As starting points, structures which closely resemble an active O-saccharide, but have variations in their conformational properties, and, or strategic functional groups modifications, are often required (*4,5*). Amongst these are acetal analogues in which one of the oxygens of the glycosidic linkage is replaced by a methylene (CH_2) residue. An additional feature of these structures is their stability to chemical and enzymatic hydrolysis.

Our interest in such methylene acetal derivatives is connected with the design of mimetics of Sialyl Lewis X (sLe^x) **1**, a proposed native ligand for the selectin family of cell adhesion molecules. sLe^x-selectin interactions have been implicated as an early phase in the inflammation process, and constitutes the basis for the design of new therapeutic agents (*6*). One approach centers on sLe^x mimetics which can act as selectin antagonists (*7*), and the 1,1-Gal-Man disaccharide **2** (*8*) has emerged as a lead compound. Although considerably less complex than sLe^x, **2** the highly substituted structure is relevant to the question of recognition specificity. In addition carbon acetal analogues of**2** are likely to have conformational properties which are quite different from the corresponding analogues of non-trehalose type disaccharides (*9, 10*). Herein we describe the synthesis of the C-glycoside **3**, and derivatives of the aza-C-glycoside **4** (*11*), and the carbaglycoside **5**. The methodology is a potentially general one for such acetal analogues of β-galacto-disaccharides (Fig. 1).

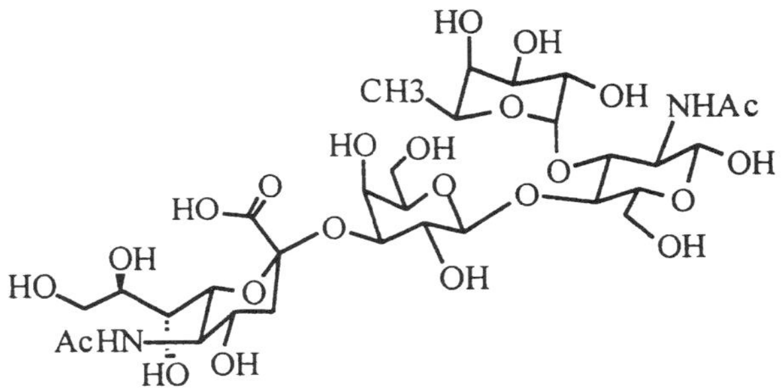

1 Sialyl Lewis X (sLe^x)

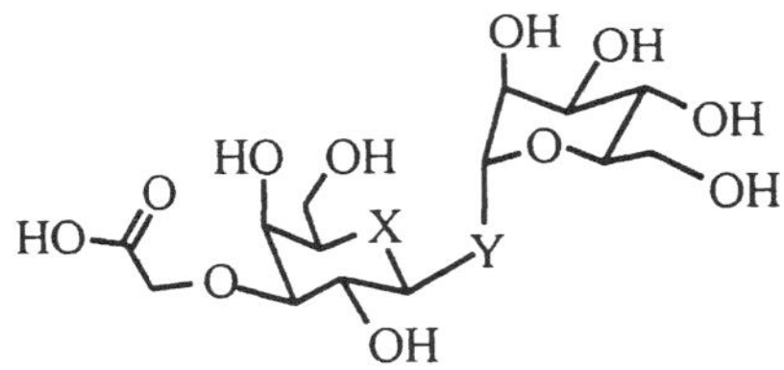

2 X = Y = O; sLe^x Mimetic
3 X = O, Y = CH_2; C-Glycoside Analog
4 X = NH, Y = CH_2; Aza-C-Glycoside Analog
5 X = CH_2, Y = O ; Carbagycoside Analog

Fig. 1: sLe^x and Acetal Analogues of sLe^x Mimetic **2**

Of the different acetal analogues, C-glycosides have received the most attention from synthetic chemists (*12-16*). "True" C-disaccharides, in which the only structural difference from the O-disaccharide is the substitution of the intersaccharide oxygen by a methylene, are generally more synthetically challenging than C-linked analogues which have longer or functionalized intersaccharide linkers, or unnatural aglycone segments. The most common strategy for C-disaccharides involves the coupling of cyclic "glycone" and "aglycone" components through the construction of the "glyconic" or "aglyconic" C-C bond (*17-22*). Practical aspects of this general approach are easy access to the individual components, and the convergent design. However, low coupling efficiency is a common problem. Strategies involving "sugar" ring formation are also known, but generally involve lengthy, linear reaction sequences (*23-25*). This approach is popular for aza-C-glycosides, since preformed piperidine precursors are not as easily accessible as their ether analogues (*26-28*). There are less practical bond disconnections for carbadisaccharides, than for the C-glycoside derivatives. One of the more general approaches entails the use of epoxysugars or epoxycyclitols and alcohol/amine partners (*29*, *30*). Two major problems are the dearth of concise routes to cyclitol components, and poor reaction yields when bond formation to a secondary carbon is involved.

1-Thio-1,2-Isopropylidene Acetals (TIA's)

Background

We envisaged assembly of analogues like **3-5** via construction of the C2-C3 bond of the galacto ring. This strategy evolved from an earlier methodology for the synthesis of adjacently linked cyclic ethers (*31*). Treatment of the bis-alkenylfuranoside **6** with iodonium dicolidine perchlorate (IDCP) led to the THF-THP product **8**. Noteworthy aspects of this reaction are the high yield and the excellent stereoselectivity in formation of the THP ring. The mechanism is believed to proceed via neighboring group participation by the ring oxygen of the furanoside onto a first formed iodonium ion. Fragmentation of the THF onium leads to the cyclic oxocarbenium ion **7**, which is attacked by the pendant alkene to give the THP ring. This subunit may be regarded as 1-deoxy-*galacto* alditol, and suggested the use of 1,2-O-isopropylidenes like **9** and **10** as annulating synthons for highly oxygenated structures (Scheme 1).

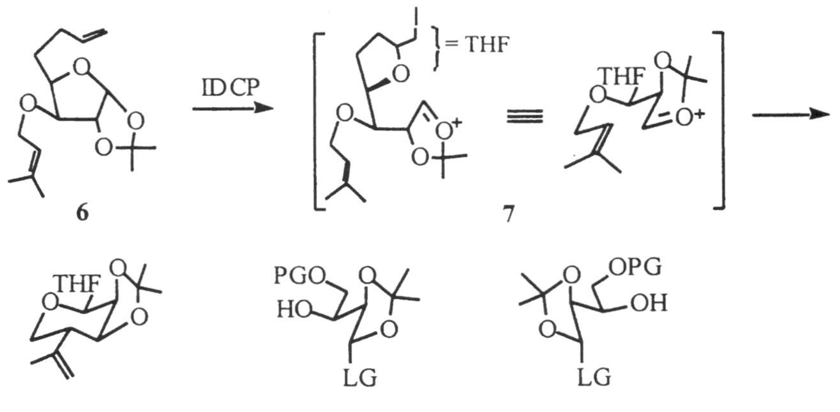

D/L *Galacto* Synthons

Scheme 1

TIA's as Glycomimetic Synthons

Methylenation of the ester derived from 1-thio-1,2-O-isopropylidene (TIA) alcohol **18** and the acid **19**, is expected to to give an enol ether-thioacetal **16** (*32*). Thioacetal activation in **16**, should provide the versatile C1-substituted galactal **14**. Subsequent elaboration of **14** could lead to the β-C- and the aza-β-C-galactoside as well as a variety of other derivatives. Juxtaposition of the alcohol and acid moieties in the initial esterification partners (i.e. **20, 21**), followed by a similar sequence should provide the corresponding β-carbagalactoside **13**. Thus, the TIA methodology could provide a general routes to C-, -aza-C- and carba- derivatives of a given β-galactoside (Scheme 2).

C-Glycosides

A key reaction in the synthesis of the TIA alcohol **18** was the Suarez fragmentation (*33*) of the 2,3-O-isopropylidene-D-lyxose derivative **23** (*34*). Treatment of **23** with diiodobenzene diacetate (DIB) provided the 1,2-O-isopropylidene acetate **24** in 88% yield. Acetal exchange was effected by treatment of **24** with thiophenol and $BF_3.Et_2O$ at -78 °C. Mild base hydrolysis of the product provided **18** in 90% yield from **24**, and in 65% overall yield over the four one-pot operations for commercially available D-lyxose **22** (Scheme 3).

For the synthesis of the C-galactoside **11**, TIA **18** was subjected to DCC mediated esterification with 1.2 equivalents of the known C-*manno* acid **19**

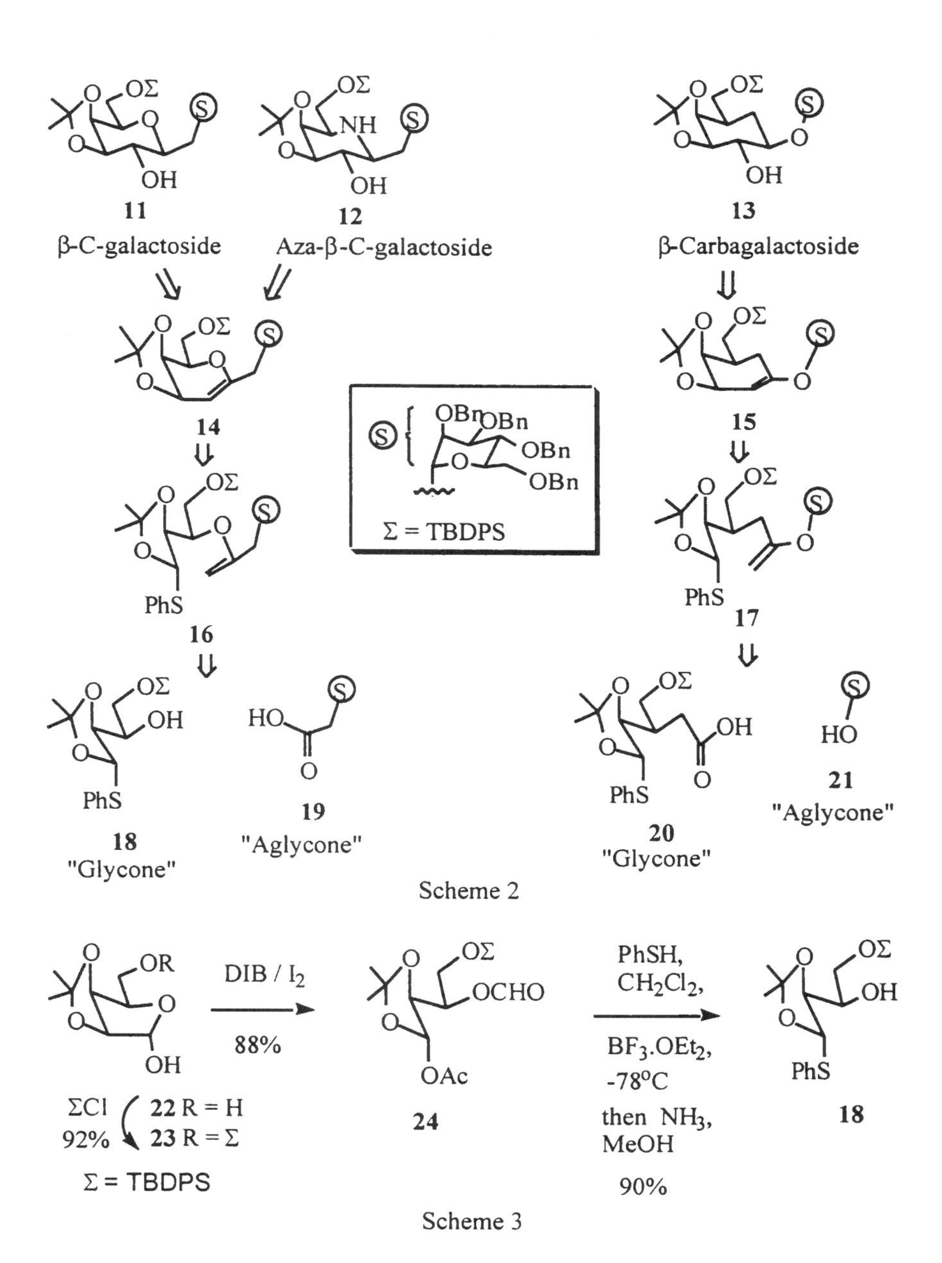

Scheme 2

Scheme 3

(*35*). Tebbe reaction on the resulting ester **25**, provided the enol ether **16** in 85% overall yield from **18** (Scheme 4). The key cyclization reaction was promoted by methyl triflate in the presence of 2,6-di-t-butyl-4-methylpyridine (DTMP), and fresly activated molecular sieves, in CH_2Cl_2. The galactal **14** was obtained in 88% yield. No evidence was observed for any of the exo-glycal isomer of **14**. Hydroboration of **14** proceeded smoothly to the β-C-galactoside **11** as a single diastereomer. The stereoselective hydroboration of related C1 substituted glycals has been reported (*25*).

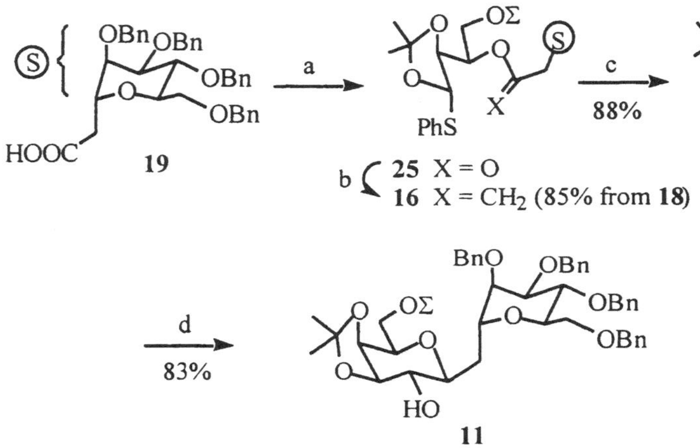

25 X = O
16 X = CH_2 (85% from **18**)

(a) **18**, DCC, DMAP, PhH; (b) Tebbe; (c) MeOTf, DTBMP, CH_2Cl_2;
(d) BH_3.DMS then Na_2O_2

Scheme 4

C-disaccharide **11** was next transformed to the target compound **3** via a straightforward sequence of reactions (*36*). Thus acetonide hydrolysis in **11** provided the 3,4,6-triol which was subjected to Bu_2SnO mediated selective alkylation with methyl bromoacetate (*37*). This reaction produced a mixture of O2 and O4 lactone derivatives of the O3 alkylated product. Basic hydrolysis of this mixture afforded, a single triol acid **26**, in 51% overall yield from **11**. Finally hydrogenolysis of **26** led to **3**, the C-glycoside analog of sLe^x mimetic **2** (Scheme 5).

NMR analysis indicated that the O-disaccharide derivative of **11**, exists primarily in the exo-syn/exo-syn conformation (93%), corresponding to the operation of two exo-anomeric effects (*38*). By comparison the C-glycoside **11** is considerably more flexible with distributions between 8-40% over five major conformer populations. The glyconic bond to the *galacto* component (exoanomeric orientation in 72% of all conformations) is more rigid than the glyconic bond to the *manno* segment (exoanomeric orientation in 48% of all conformations). Comparison of the binding affinity of**2** and **3** might therefore shed light on issues of conformational rigidity and induced fit, as they pertain to the sLe^x-selectin recognition.

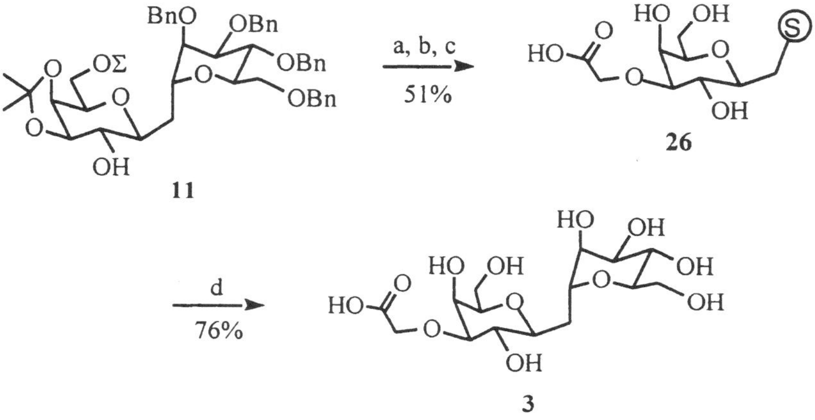

(a) MeOH, HCl; (b) Bu_2SnO then $BrCH_2CO_2Me$;
(c) KOH, EtOH-H_2O, then H^+; (d) H_2, $Pd(OH)_2/C$, MeOH

Scheme 5

The C-glycosidation methodology was also applied to the 1-6 and 1-4 linked C-disaccharides **29** and **32**, the C-monosaccharide **35** and the C-glycoside containing an aminated aglycone, **38** (*32*). In the latter case, the sulfonamide was found to be a compatible amine protecting group. The synthesis of these structures paves the way for application to a variety of C-disaccharides, C-glycolipids and C-glycosyl amino acids of biological significance (Table 1).

The strategy is also applicable to different glycone analogues, as illustrated in the synthesis of the furano-glycal **44**, a potential precursor to derivatives of 3-deoxy-D-*manno*-octulosonic acid (KDO) (*39*). Thus, 2,3:5,6-di-O-isopropylidene-*manno*-furanose **39** was subjected to the Suarez fragmentation, and the product **40** was converted to the TIA alcohol **41** via a similar sequence as outlined for TIA **18**. Esterification of **41** with 2-furoic acid, followed by Tebbe methylenation of the ester **42** provided the enol ether **43**. Treatment of **43** under the standard cyclization conditions led to te glycal **44** in 85% yield (Scheme 6).

Table 1: C-Glycosides Prepared from TIA 18

Acid (HOOC–Ⓢ)	Glycal (%Yield from **18**)	C-Glycoside (%Yield)
19 (Ⓢ = tetra-O-benzyl sugar: OBn, OBn, OBn, OBn)	**14** (65)	**11** (75)
27 (di-O-isopropylidene sugar)	**28** (43)	**29** (75)
30 (OBn, OBn, OBn, OMe)	**31** (60)	**32** (77)
$CH_3(CH_2)_{15}$ **33**	**34** (80)	**35** (80)
$(CH_2)_4$–N(SO_2Ph)(Bn) **36**	**37** (50)	**38** (84)

a 10% of exo-glycal isomer obtained, easily converted to **28** on heating in benzene

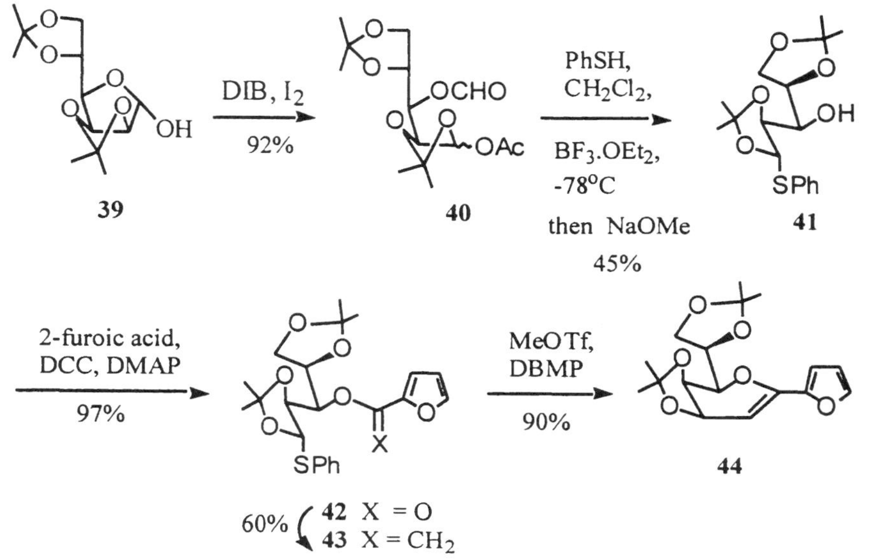

DIB, I_2
92%

OCHO
OAc
OH

39 40

PhSH, CH_2Cl_2, $BF_3.OEt_2$, -78°C then NaOMe 45%

OH
SPh
41

2-furoic acid, DCC, DMAP
97%

SPh
X

MeOTf, DBMP
90%

44

60% (**42** X = O; **43** X = CH_2)

Scheme 6

Aza-C-galactosides

We envisaged access to the aza-C-galactoside **12** via a stereoselective double reductive amination on a diketone precursor. Previous results from this laboratory and others suggested that the two new stereogenic centers created in this reaction might proceed through the reduction of cyclic iminium ions, and should favor the desired β-galacto type motif (*27*, *40*). The glycal **14** was considered as a precursor to a suitable diketone derivative (Scheme 1).

Accordingly, **14** was converted under standard dihydroxylation conditions to hemiketal **45** in 80% yield. Selective benzylation of **45** provided benzyl ether **46**, which was transformed to the diketone **47**, over two steps: ketone reduction, and subsequent reoxidation of the resultant diol. Treatment of **47** under our standard reductive amination conditions (1.2 equivalents NH_4HCO_2, $NaCNBH_3$, dry MeOH, 4A MS, 2h) provided the aza-β-C-galactoside **12** in 72% yield. The structure of **12** was confirmed by HCOSY, ^{13}CNMR and MS analysis. No diastereomeric products were observed within the limits of NMR detection (Scheme 7).

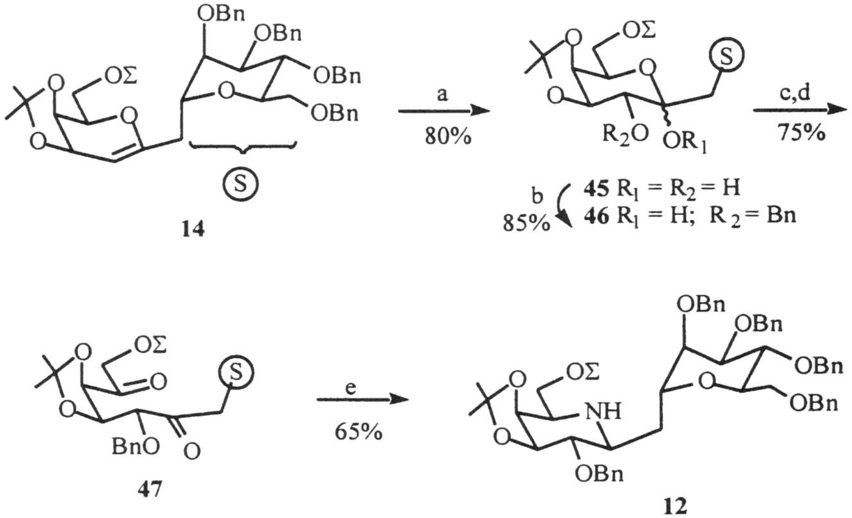

(a) OsO_4, NMNO, acetone; (b) BnBr, AgOTf, CH_2Cl_2; (c) $NaBH_4$;
(d) Swern Ox.; (e) 1.3 eq. NH_4HCO_2, 2 eq. $NaCNBH_3$, dry MeOH, 4A MS

Scheme 7

Carbagalactosides

The TIA acid **50** corresponding to the glycone component for the carbadisaccharide synthesis, was obtained as an inseparable mixture with the epimer **51**. The starting material for **50/51** was the TIA alcohol **18** which was used in the C-glycoside methodology. In view of subsequent reactions it was necessary to first replace the silyl ether in **18** with a benzyl protecting group. Swern's oxidation of **48**, followed by treatment of the resulting ketone with carbomethoxymethylene triphenylphosphorane afforded **49** as a single alkene isomer of undetermined stereochemistry. Hydrogenation of **49** over palladium on carbon provided a 1:1 mixture of **50/51**. The stereochemistry of these products was assigned subsequently in the carbadisaccharide products. The (R,R,-Me-DUPHOS)-Rh catalysed hydrogenation (*41*) gave a more favorable selectivity for the desired isomer (**50/51** = 8:1). However, this catalyst is apparently easily deactivated, and these hydrogenations sometimes proceeded with low substrate conversion.

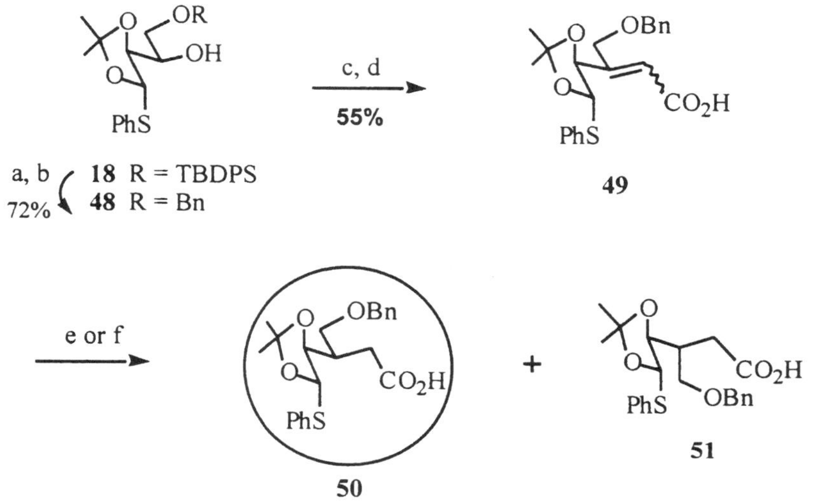

(a) nBu_4NF, THF, 97%; (b) Bu_2SnO, BnBr, PhH, 95%; (c) Swern Ox. 88%; (d) $NaCHCO_2EtPO(OEt)_2$ then aq. NaOH, 68%; (e) Pd/C, H_2, 1 atm; 99%, **50:51** = 1:1; (f) 5%[$Rh(COD)_2OTf$ / (R,R)-Me-DuPhos, MeOH, 55 psi, 4 days, 90% based on recovered **49**; **50:51** = 8:1

Scheme 8

In the event, it was possible to carry through the mixture of **50/51**, through subsequent steps and separate the diasteromers products at a later stage. Accordingly DCC mediated esterification of a 1:1 mixture of **50/51** with the mannose derivative **52**, led to a mixture of ester products **53**, which under Tebbe reaction conditions provided a mixture of the enol ethers, **54**. Treatment of **54** under the standard cyclization conditions afforded a mixture of cyclic enol ethers **55** in 74% yield. Hydroboration of this mixture gave a separable mixture of**56** and **57** in a combined yield of 82%. The identity of these products was confirmed by HCOSY, ^{13}CNMR and MS analysis. The absence of any other products in the hydroboration reaction indicated that the deprotonation step in the key cyclization step was, as in the C-glycoside systems, highly regioselective. It is likely that this selectivity arises from torsional factors associated with the *cis*-isopropylidenoxy residue.

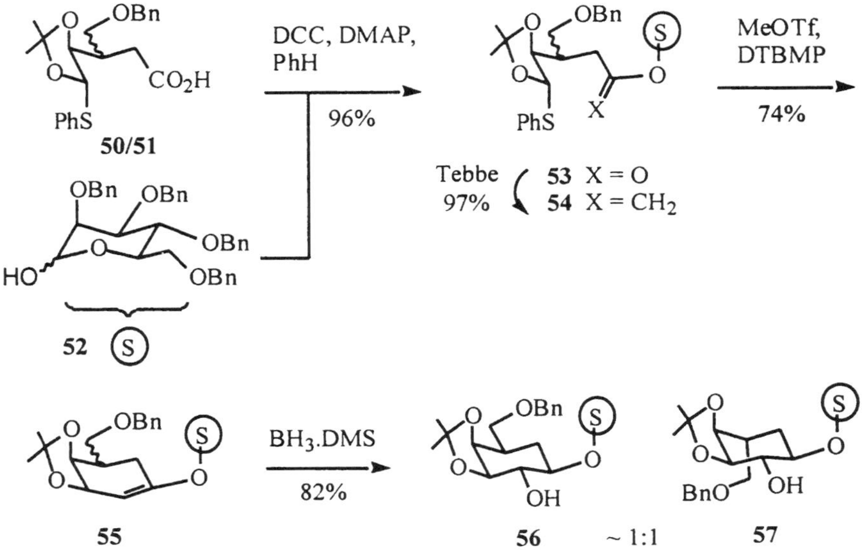

Scheme 9

Summary

The foregoing chemistry constitutes a potentially general protocol for preparing the C-glycoside, aza-C-glycoside and carbasugar analogues of a given β-galactoside from a single TIA precursor **18**. Important aspects of these syntheses are the efficiency of the coupling with complex "aglycone" components, and the high stereoselectivity at the pseudoanomeric position. As illustrated for the analogues of the sLe^{x} mimetic **2**, these attributes are especially relevant to disaccharide systems. This chemistry also lays the groundwork for applications of TIA chemistry to other classes of complex cyclic ethers and cyclohexanes.

Acknowledgment

This investigation was supported by NIH grant GM 57865. "Research Centers in Minority Institutions" award RR-03037 from the National Center for Research Resources of the NIH, which supports the infrastructure {and instrumentation} of the Chemistry Department at Hunter, is also acknowledged. The contents are solely the responsibility of the authors and do not necessarily represent the official views of the NCRR/NIH.

References

1. Sears, P.; Wong, C.-H. *Angew. Chem. Int. Ed. Eng.* **1999**, *38*, 2300 2324.
2. *Carbohydrates in Drug Design*; Witczak, Z.J., Nieforth, K.A., Eds.; Marcel Dekker, Inc.: New York, NY, 1997.
3. *Carbohydrate Mimics: Concepts and Methods*; Chapleur, Y. Ed.; Wiley-VCH.: New York, NY, 1998.
4. Lemieux, R.U.; Du, M.-H.; Spohr, U. *J. Am. Chem. Soc.* **1994**, *116*, 9803-9804.
5. Bundle, D.R.; Alibés, R.; Nilar, S.; Otter, A.; Warwas, M.; Zhang, P. *J. Am. Chem. Soc.* **1998**, *120*, 5317-5318.
6. Boschelli, D.H. *Drugs of the Future* **1995**, *20*, 805-816.
7. Simanek, E.E.; McGarvey, G.J.; Jablonowski, J.A.; Wong, C.-H. *Chem. Rev.* **1998**, *98*, 833-862.
8. Shibata, K.; Hiruma, K.; Kanie, O.; Wong, C.-H. *J. Org. Chem.* **2000**, *65*, 2393-2398.
9. Ravishankar, R.; Surolia, A.; Vijayan, M.; Lim, S.; Kishi, Y. *J. Am. Chem. Soc.* **1998**,*120*, 11297-11303.
10. Espinosa, J.F.; Montero, E.; Vian, A.; Dietrich, H.; Schmidt, R.R.; Martín-Lomas, M.; Imberty, A.; Cañada, F.J.; Jiménez-Barbero, J. *J. Am. Chem. Soc.* **1998**, *120*, 1309-1318.
11. In this manuscript imino alditols such as **4** are referred to by the commonly used, albeit incorrect name, "azasugar".
12. Postema, M.H.D.*C-Glycoside Synthesis*, CRC Press, Boca Raton, Fl, 1995.
13. Levy, D.; Tang, C. *The Synthesis of C-Glycosides*, Pergamon, Oxford, 1995.
14. Beau, J.-M.; Gallagher, T. *Topics Curr. Chem.* **1997**, *187*, 1-54.
15. Togo, H.; He, W.; Waki, Y.; Yokoyama, M. *Synlett* **1998**, 700-717.
16. Du, Y.; Linhardt, R.J.; Vlahov, J.R. *Tetrahedron* **1998**, *54*, 9913 9959.
17. Recent syntheses of C-disaccharides: Witczak, Z.J.; Chhabra, R.; Chojnacki, J. *Tetrahedron* **1997**, *38*, 2215-2218, and ref 18-25.
18. Rekai, El-D.; Rubinstenn, G.; Mallet, J.-M.; Sinaÿ, P. *Synlett* **1998**, 831-834.
19. Zhu, H.; Vogel, P. *Tetrahedron Lett.* **1998**, *39*, 31-34.

20. Dondoni, A.; Kleban, M.; Zuurmond, H.; Marra, A. *Tetrahedron Lett.* **1998**, *39*, 7991-7994.
21. Bazin, H.G. Du, Y.; Polat, T.; Linhardt, R.J. *J. Org. Chem.* **1999**, *64*, 7254-7259.
22. Zhu, Y.-H.; Demange, R.; Vogel, P. *Tetrahedron: Asymmetry* **2000**, *11*, 263-282.
23. Sutherlin, D.P.; Armstrong, R.W. *J. Org. Chem.* **1997**, *62*, 5267-5283.
24. Pham Huu, D.-P.; Petrusova, M.; BeMiller, J.N.; Petrus, L. *Tetrahedron Lett.* **1999**, *40*, 3053-3056.
25. Postema, M.H.D.; Calimente, D. *Tetrahedron Lett.* **1999**,*40*, 4255-4259.
26. Johnson, C.R.; Miller, M.W.; Golebiowski, A.; Sundram, H.; Ksebati, M.B. *Tetrahedron Lett.* **1994**, *35*, 8991-8994.
27. Martin, O.R.; Liu, L.; Yang, F. *Tetrahedron Lett.* **1996**, *37*, 1991-1994.
28. Frérot, E.; Marquis, C.; Vogel, P. *Tetrahedron Lett.* **1996**, *37*, 2023-2026.
29. Suami, T.; Ogawa, S. *Adv. Carbohydr. Chem. Biochem.* **1990**, *48*, 22-90.
30. Ogawa, S.; Hirai, K.; Yamazaki, T.; Nakajima, A.; Matsunaga, N.; *Carbohydr. Lett.* **1996**, *2*, 183-188.
31. Khan, N.; Xiao, H.; Zhang, B.; Cheng, X.; Mootoo, D.R. *Tetrahedron* **1999**, *55*, 8303-8312.
32. Khan, N.; Cheng, X.; Mootoo, D.R. *J. Am. Chem. Soc.* **1999**, *121*, 4918-4919.
33. De Armas, P.; Francisco, C.G.; Suarez, E. *Angew. Chem. Int. Ed. Engl.* **1992**, *31*, 772-774.
34. Levene, P.A.; Tipson, R.S. *J. Biol. Chem.* **1936**, *115*, 731-747.
35. Wong, C.-H.; Moris-Varas, F.; Hung, S.-C.; Marron, T.G.; Lin, C.-C.; Gong, K.W.; Weitz-Schmidt, G. *J. Am. Chem. Soc.* **1997**, *119*, 8152-8158.
36. Cheng, X.; Khan, N.; Mootoo, D.R. *J. Org. Chem.* **2000**, *65*, 2544-2547.
37. David, S.; Hanessian, S. *Tetrahedron* **1985**, *41*, 643-663.
38. Asensio, J.L.; Cañada, F.J.; Cheng, X.; Khan, N.; Mootoo, D.R.; Jiménez-Barbero, J. *Chem. Eur. J.* **2000**, *6*, 1035-1041.
39. Uner, F.M. *Adv. Carbohydr. Chem. Biochem.* **1981**, *38*, 323-.
40. Zhao, H.; Mootoo, D.R.; *J. Org. Chem.* **1996**, *61*, 6762-6763.
41. Burk, M.J.; Gross, M.P.; Martinez, J.P. *J. Am. Chem. Soc.* **1995**, *117*, 9375-937

Chapter 7

Iminosugars, Isoiminosugars, and Carbasugars from Activated Carbohydrate Lactones: Efficient Synthesis of Biologically Important Compounds

Inge Lundt

Technical University of Denmark, Department of Organic Chemistry, Building 201, DK–2800 Lyngby, Demark

The synthetic potential of selectively activated aldonolactones as building blocks for synthesis of optically pure, highly functionalised organic molecules is highlighted. Preparation of iminosugars from selectively brominated lactones requires only two transformations, in which the ring closure by reaction with ammonia is the key step. Stereoselective alkylation of unprotected bromodeoxylactones offers a general synthetic approach to isoiminosugars. Radical induced carbocyclisation of ω-bromo-α,β-unsaturated aldonolactones yields functionalised cyclopentane /cyclohexane lactones stereospecifically, generating one or two chiral centers in the ring closing step. Stereo- and regio-selective functional group interconvertion within the bicyclic cyclopentane-lactone system gives access to hydroxy/amino substituted cyclopentanes. Aldonolactones provide thus in few steps access to compounds of biologically importance in an optically pure state.

Carbohydrate Lactones/Aldonolactones

Preparation and Chemical Reactivity

In the past twenty years aldonolactones have found widespread application as cheap, chiral synthons for the synthesis of many biologically important compounds and natural products (*1*). Particularly the area of aza- and carba-sugar synthesis has seen aldonolactones emerging as versatile starting materials (*1,2*). Aldonolactones constitute a more diverse chiral pool of compounds than aldoses. From each aldose several aldonolactones/aldonic acids are available in just one step, *i.e.* by anomeric oxidation of the aldose (*3*), by one carbon Kiliani chain elongation [4] or by one carbon oxidative degradation (*5*). In addition, a number of aldonolactones are available by some more special reactions including reduction, or oxidative cleavage of the double bond in vitamin C (*6*). As a result, aldonolactones are in many cases more readily available than the corresponding aldoses, especially in the L-series.

The chemical reactivity of aldonolactones also differ remarkably from the reactivity of aldoses. For synthetic manipulation of aldoses several steps are usually required at the outset to protect and define the stereochemistry at the C-1 hemiacetal function *e.g.* via a glycoside. The lactone group at C-1 in aldonolactones, however, can be maintained through a number of transformations (*1,2*). The reactivity of the hydroxy groups in aldonolactones is also different from what is normally observed in aldoses. Particularly the hydroxy group α to the lactone group show similar or enhanced reactivity as compared to the primary hydroxy group. This gives rise to a number of regioselective reactions in aldonolactones and diminishes the need for many different protecting groups (*1,2*). Furthermore, it should be noted that aldonolactones usually prefer the 5-membered 1,4-lactone form contrary to the 6-membered pyranose form predominant in aldoses.

Preparation of Activated Aldonolactones

The regioselective functionalisation of aldonolactones is possible at the position α to the lactone and at the primary (ω) position. The most efficient method for this functionalisation is treatment of the aldonolactone/aldonic acid with hydrogen bromide in acetic acid (Scheme 1). In this strongly acidic medium the lactone is partly acetylated followed by formation of acetoxonium ions. These then undergo opening with bromide ions to give acetylated bromodeoxyaldonolactones (*7*). The formation of acetoxonium ions controls the regio- and stereo-selectivity of introduction of the bromine: bromide is always opening the acetoxonium ion at the primary position and at C-2, with inversion of the configuration. These are formed from 1,4-lactones having the OH-2 and OH-3 in a *cis*-orientation. In two cases however, in glucono- and xylono-lactone having the OH-2 and OH-3 in a *trans*-orientation, bromine is also introduced at C-2 with inversion of the configuration. The acetoxonium ions are here formed between two *trans* hydroxy groups from the open aldonic acids, which in these cases are present to a certain extend. In general, the aldonolactones adopt the 1,4-lactone form (*7*).

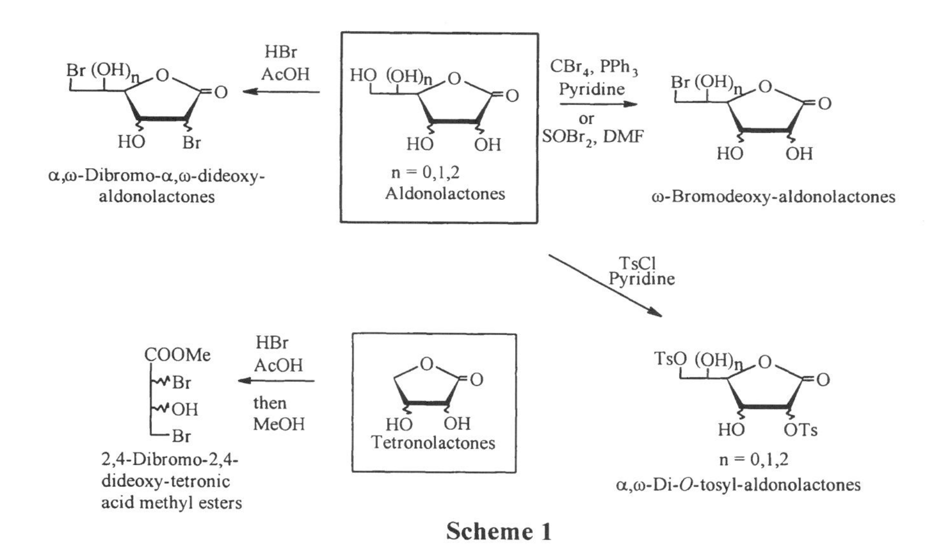

Scheme 1

Carbohydrate Synthons: Preparation of selectively activated aldonolactones

Regioselective bromination of the primary position can also be achieved with carbontetrabromide and triphenylphosphine in pyridine (*8*) or thionyl bromide in DMF (*9*). Regioselective mesylation or tosylation of the primary hydroxy group, however, is inefficient due to competing sulfonylation of theα-hydroxy group. Instead, α,ω-di-*O*-tosylated aldonolactones can in some cases be formed in good yields by using 2.0-2.3 eq. of tosyl chloride in pyridine (*10*). The 2-*O*-tosylated aldonolactones thus formed have *the same configuration as the starting lactones and thus are C-2 epimers to the bromodeoxy aldonolactones prepared by with hydrogen bromide/acetic acid.* Access to new activated lactones has thus been achieved.

Carbohydrate Mimics

Development of carbohydrate-based therapeutics has been hampered by the enzymatic biodegradability of these molecules. Carbohydrates, as part of glycoproteins, glycolipids and other glycoconjugates, are essential partners in the important processes such as cell-cell communication, and molecular and cellular targeting. This insight, although not fully understood, has led to the discovery that interfering with the appropriate enzymatic processes by use of *enzyme inhibitors*, *carbohydrate mimics*, may have valuable therapeutic effects and contribute to the understanding of such fundamental processes. The term "*carbohydrate mimics*" is used for small molecules that contain essential functional groups to resemble the structure and the conformation of the parent carbohydrate in the transition state of the enzymatic processes. Several sugar analogues with basic nitrogen instead of oxygen in the ring (iminosugars), or with this oxygen beeing replaced by a methylene group, are stable towards hydrolysis due to the lack of the acetal function. These type of compounds have been recognised as glycosidase inhibitors.

Iminosugars

Synthetic approaches

Preparation of iminosugars from activated aldonolactones requires two transformations: ring closure to form the pyrrolidine/piperidine ring and reduction of the lactone group. To carry out these transformations we have developed four different strategies (*2*) as shown in Scheme 2. Strategy I relies on direct ring closure of the activated lactones with ammonia to form iminoamides, which subsequently are reduced to iminoalditols. In strategy II the two steps have been interconverted. Strategy III implies reaction of activate

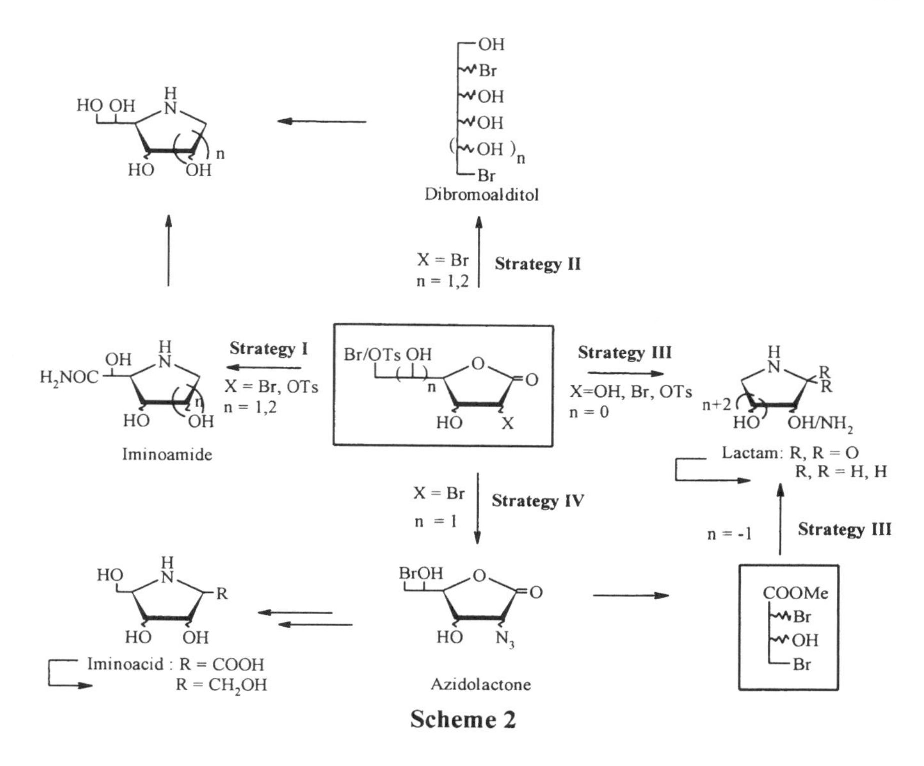

Scheme 2

Strategies for preparation of iminosugars from selectively activated aldonolactones

pentono- or tetrono lactones/esters with ammonia to give lactams, which are subsequently reduced to the iminoalditols. Strategy IV differs from the other strategies by introducing the nitrogen *via* an azide. Diactivated lactones can selectively be transformed to 2-azido lactones, which by azide reduction and ring closure give rise to iminoacids. Reduction then affords pyrrolidines.

The ring closure of diactivated aldonolactones by ammonia is a very easy method to obtain iminoalditols. By monitoring the reactions by ^{13}C NMR spectroscopy, it has been shown that α,ω-diactivated aldonolactones react with the basic aqueous ammonia to give diepoxy aldonic acids, having an epoxy group at the α,β-position and at the primary position. The primary epoxide, as the more reactive of the epoxides formed, was subsequently opened by ammonia to give a ω-amino-ω-deoxy-α,β-epoxy-aldonic acid. Subsequent attack by the primary amino group at C-3 (sugar numbering) of the remaining epoxy group yielded the heterocyclic ring, as exemplified in Scheme 3 (*2, 11a,12*).

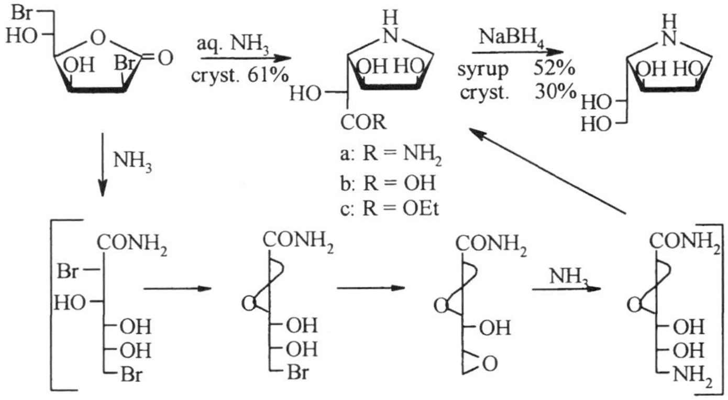

Scheme 3

5-Membered iminosugars from bromodeoxyhexonolactones

Reaction of C-5 activated pentonolactones (*13*) and of 2,7-diactivated heptonolactones (*11b,13*) with ammonia gave in both cases 6-membered imino sugars by similar formation of epoxide intermediates (Scheme 4). The 6-bromo-2,6-dideoxy-D-*arabino*-hexono-1,4-lactone gave by reaction with ammonia the 7-membered lactam (*14*) (Scheme 4).

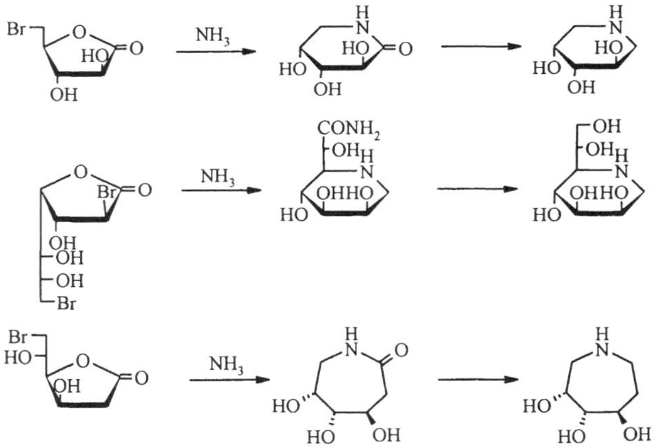

Scheme 4

6- and 7-membered iminosugars from bromodeoxyaldonolactones

Using this type of chemistry, more than 30 different iminoalditols have been synthesised without using any protecting group strategy, and a number of new glycosidase inhibitors have been identified (*2,12,13,14*). Some selected interesting features will be discussed below.

Glycosidase inhibitory properties

The cleavage of glycosidic bonds is catalysed by acids, and the enzymatic cleavage is catalysed by amino acid residues in the active site of the enzymes. Two different mechanisms may operate, resulting in either retention or inversion of the anomeric configuration (*15*). During the hydrolytic step the partial positive charge being developed at C-1 of the sugar will be stabilised as an oxocarbenium ion, resulting in a double bond character between the ring oxygen and the anomeric carbon. This cause distortion of the conformation of the pyranose ring to a half chair (*15*). The sugar mimics have thus often been found among the 5-membered iminosugars, since they posses a planar structure

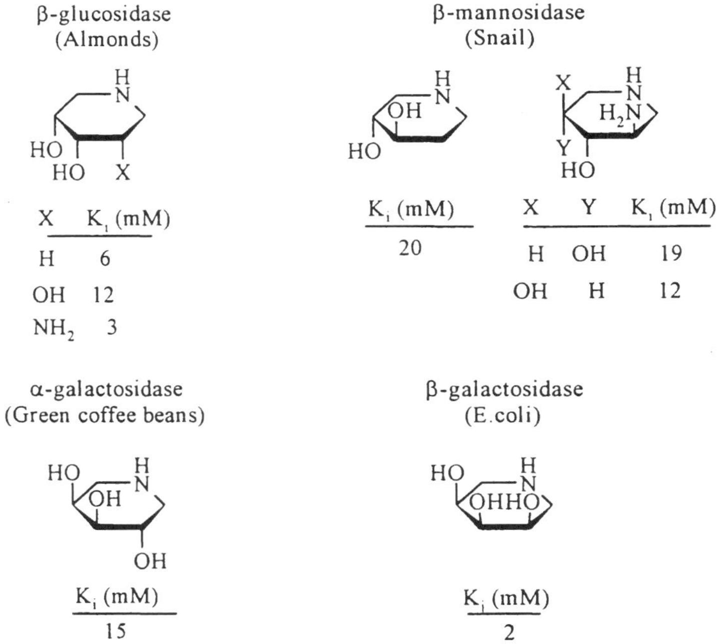

β-glucosidase
(Almonds)

X	K_i (mM)
H	6
OH	12
NH_2	3

β-mannosidase
(Snail)

K_i (mM)
20

X	Y	K_i (mM)
H	OH	19
OH	H	12

α-galactosidase
(Green coffee beans)

K_i (mM)
15

β-galactosidase
(E.coli)

K_i (mM)
2

Scheme 5

6-Membered 5-carbon amino/hydroxy iminosugars as inhibitors of hexosidases having the nitrogen, protonated by the amino acid residues, and three other carbons of the ring within a plane. The structure-activity-relationship (SAR) is often not fully understood, and different ring size iminoalditols as well as bicyclic structures have been identified as valuable enzyme inhibitors (*16*).

Among the 6-membered 5-carbon iminoalditols, which we have synthesised, we have interestingly found good inhibitors for glycosidases of hexoses in spite of their lacking hydroxymethyl group (*2,13*). Thus, β-glucosidase inhibitors as well as inhibitors of β-mannosidase, of which not many simple molecules are known, have been synthesised (Scheme 5). The 1,5-imino-pentitol having L-*arabino*-configuration was identified as a good α-galactosidase inhibitor. It can be viewed as galactostatin lacking the C-6 hydroxymethyl group. Galactostatin is the parent 6-membered "iminogalactose" which is the "natural" α-galactosidase inhibitor. The iminopentitol, having L-*ribo*-configuration and likewise lacking the C-6 hydroxymethyl group, inhibits β-galactosidase. Both compounds do have two *cis*-hydroxy groups at positions, corresponding to C-3 and C-4 in galactose. This structural motif may be responsible for their inhibitory properties found.

Most interesting is the finding, that three stereoisomeric 6-membered 7-carbon iminoalditols, which were synthesised from activated heptonolactones by reaction with ammonia (*11b*), were found to be powerful inhibitors towards α-L-fucosidase from human liver enzymes (*13*) (Scheme 6). The known 1,5-dideoxy-1,5-imino-D-arabinitol (*17,18*) which we have synthesised more conveniently from 5-bromo-5-deoxy-D-arabinono-1,4-lactone by stratgy II (Schemes 2 and 4) (*13*), also inhibits α-L-fucosidase. The minimal structural motif for this inhibition is the absolute configuration of the hydroxy bearing carbon atoms, which should be the D-*arabino* in 6-membered iminoalditols (*19*). The 7-carbon 1,5-iminoalditols are stronger inhibitors (Ki 16, 9 and 3 μM, Bovine kidney) compared to the parent 1,5-imino-D-arabinitol (Ki 30 μM), showing the importance of a carbon branching at the piperidine ring (Scheme 6).

Finally, we have synthesised some trihydroxylated 7-membered iminoalditols (*14*). We hoped that due to the flexibility of the azepane ring system, certain conformational advantages over the 5- and 6-membered ring inhibitors in terms of fitting into the active site of glycosidases, might emerge. Recently, a tetrahydroxyazepan with C_2-symmetry was found to exhibit noteworthy glycosidase inhibitory properties against quite a range of different types of glycosidases (*20*). The 1,2,6-trideoxy-1,6-imino-D-*arabino*-hexitol (Scheme 4) inhibited α-L-fucosidase with Ki 65 μM (Scheme 6). The inhibition was in the same range as for the 5-membered arabinitol and by superposition of the energy minimised conformations of these two compounds, good fittings for the hydroxy groups were found (*14*). The methyl group in deoxyfuconojirimycin, the most potent inhibitor of α-L-fucosidase reported to day (21) (Scheme 6), has been shown to be the vital prerequisite for this pronounced inhibition.

deoxyfuconojirimycin
Ki (μM) 0.005 (bovine epididymous)

1,5-imino-D-arabinitol

Ki (μM) 30

R_1	R_2	R_3	Ki (μM)
H	H	H	30
OH / OH	H	H	16
H	HO / HO	H	9
H	H	HO / HO	3

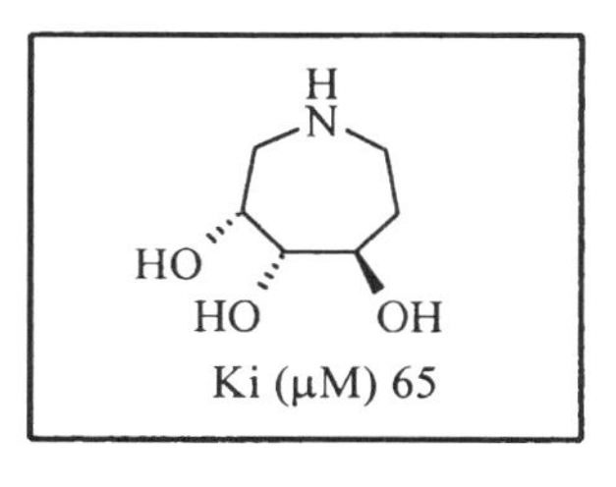

Ki (μM) 65

Scheme 6

α-L-Fucosidase inhibitors, Ki μM (Bovine kidney)

The 1,2,6-trideoxy-1,6-imino-D-*arabino*-hexitol was found to inhibit α-D-glucosidase (Ki 12 μM, Bakers yeast), β-D-glucosidase (Ki 51 μM, Almonds) and β-D-galactosidase (Ki 13 μM, E. coli) as well, indicating the flexibility of the 7-membered ring system (*14*).

Evaluation of nine 1,4-dideoxy-1,4-imino-hexitols, prepared by strategies I and II, are reported elsewhere (*2,12*). Among other results, some interesting (*22*) and selective (*12*) α-D-mannosidases were found.

Isoiminosugars/1-*N*-iminosugars

The success of iminosugars as glycosidase inhibitors has been largely attributed to their resemblance to the transition state for glycosidic cleavage as discussed above (*15, 23*). During the hydrolytic reaction, the developing positive charge at C-1 is partly decolalised by the ring oxygen. The protonated iminoalditols, mimicing the transition state, possesses glycosidase inhibitory activity, as discussed above. Changing the position of the nitrogen to the anomeric position might thus result in compounds beeing operativ as glycosidase inhibitors since part of the positive charge in the hydrolytic step will be developed at this place. These type of compounds are named isoiminosugars or 1-*N*-iminosugars.

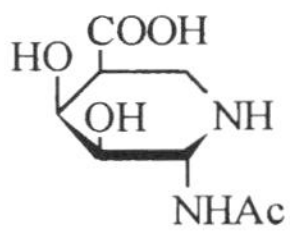

HO COOH OH NH NHAc

Siastatin B
isolated from a microorganism 1974

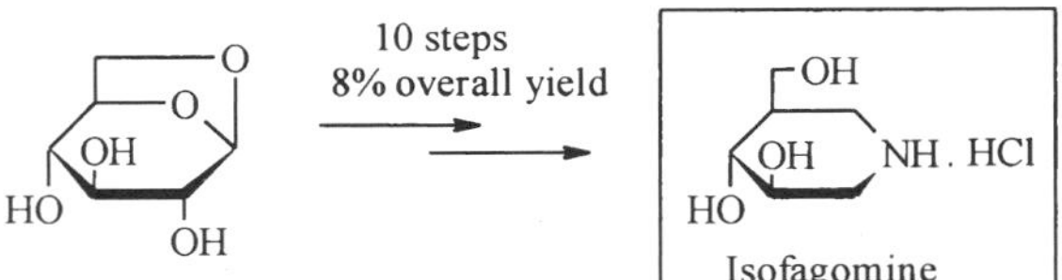

O O OH HO OH

10 steps
8% overall yield

OH OH NH . HCl HO

Isofagomine

Scheme 7

Isoiminosugars

The first reported isoiminosugar was siastatin B **1**, isolated from a microorganism in 1974 and shown to be a strong inhibitor of *N*-acetylneuraminidase (*24*). Many analogues of siastatin B **1** were prepared and tested as glycosidase inhibitors, but not untill the first synthesis of isofagomin (*25*) the full potential of isoiminosugars was realised. Isofagomin was found to be the most potent β-D-glucosidase inhibitor to date (Ki 0.11 μM, almonds) (Scheme 7). Since then many different isoiminosugars have been prepared and many have shown remarkable inhibition ofβ-glycosidases. Placing the nitrogen at the anomeric position thus seems greatly to enhance inhibition of β-glycosidases in favour of α-glycosidases (*26*), which are preferentially inhibited by iminosugars.

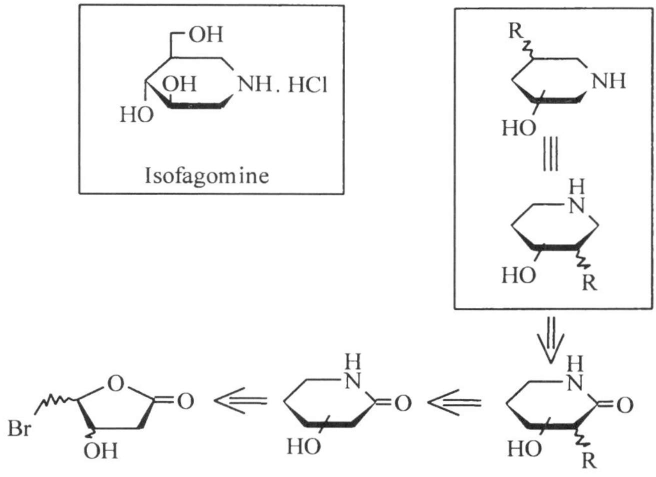

Isofagomine

Scheme 8

Strategy for the synthesis of isoiminosugars

The potential of isoiminosugars requires efficient synthetic procedures for this type of compounds, procedures which also will allow for synthesis of analogous structures. The first reported preparation of isofagomin, which was based on 1,6-anhydro-D-glucose (laevoglucosan), gave about 8% overall yield in 10 steps (*25*). Retroanalysis of the isoimino-structures reveals that they can be viewed as hydroxylated 3-alkylated piperidine derivatives, which might be obtained from α-alkylated lactams, again obtained from alkylation of 2-deoxy aldonolactams (Scheme 8). As shown above (Scheme 4) such lactams are obtainable from ω-bromodeoxyaldonolactons by reaction with ammonia. An investigation of alkylation of unprotected 5-bromo-2,5-dideoxyaldonolactons was thus initiated.

Alkylation of 5-Bromo-2,5-Dideoxypentonolactons – Synthesis of Isoiminosugars

Synthesis of 3-alkylated piperidine derivatives could of course be performed most directly by alkylation of 2-deoxyaldono-1,5-lactams, as outlined in Scheme 8. After unsucessfull experiments due to solubility problems, we turned our interest to the alkylation at C-2 of unprotected 2-deoxy-ω-bromodeoxyaldonolactons. This strategy requires set-up of a dianion in the presence of a primary bromine, followed by stereoselective alkylation at C-2, two possible doubtful reactions (Scheme 9). Gratifyingly, both reactions could be performed satisfactory.

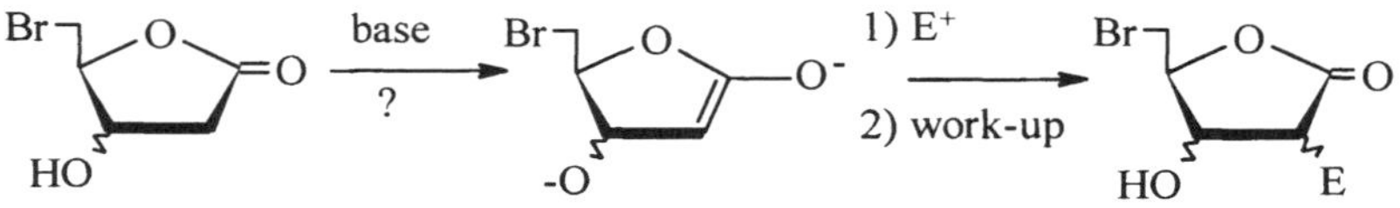

Br O =O HO base ? Br O O⁻ -O 1) E⁺ 2) work-up Br O =O HO E

Scheme 9

Alkylation of unprotected 5-bromo-2,5-dideoxy-pentono-1,4-lactones

Treatment of the 2-deoxyaldonolactone with BuLi or LDA in tetrahydrofurane at –78 °C gave the dianion within 10 min. Addition of methyl iodide as the electrophile gave after work up the 2-methylated aldonolactone, having the methyl group trans to the 3-OH group. The isolated yield was 63 % and the stereoselectivety 95:5 (*27*) (Scheme 10). A similar result was obtained from the C-3 isomeric aldonolactone (*27*). The synthesis of isofagomine by this method requires an electrophile, which gives a hydroxymethyl group at C-2, either directly or after modification.

We thus investigated a range of such one-carbon electrophiles (Scheme 10). Formaldehyd, as the most reactive one, gave the desired product directly (*28*). Using benzaldehyde as the electrophile likewise lead to a C-2 branched aldonolactone in a high yield and stereospecificity. The two diastereoisomeric compounds obtained, due to the new chiral center formed in the side chain, were both isolated in a crystalline state (*28*).

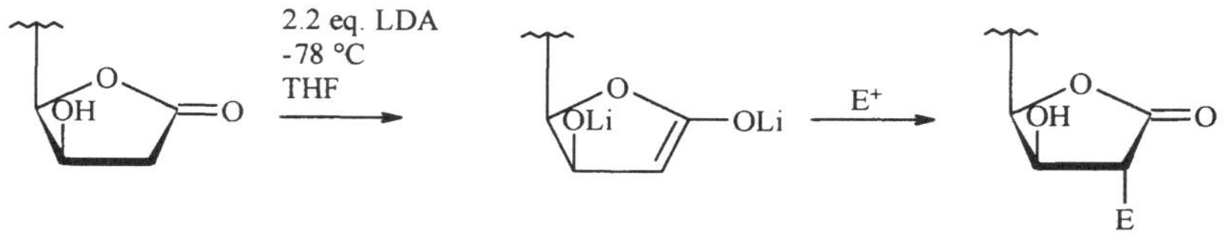

2.2 eq. LDA
-78 °C
THF

E^+

		Electrofile	E	Result
Methyliodid	:	CH_3I	CH_3	+
Methylchloroformat	:	ClCOOMe	COOMe	-
Carbondioxid	:	CO_2	COOH	-
Formaldehyde	:	CH_2O	CH_2OH	+
N,N-dimethylformamide	:	$(CH_3)_2NCHO$	CHO	-
Ethylformiate	:	HCOOEt	=CHOH	-

Scheme 10

Alkylation using electrophiles which might give a one-carbon branching at C-2

Having the 2-alkylated 5-bromolactones in hand, the conversion to isoiminosugars were performed by transforming the bromine to an amino group *via* an azide displacement followed by reduction. This gave directly the optically pure lactams by crystallisation. Reduction of the lactam function gave

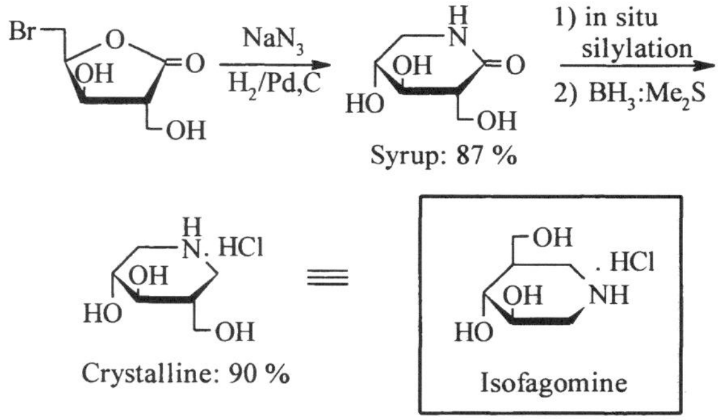

NaN_3
$H_2/Pd,C$

Syrup: 87 %

1) in situ silylation
2) $BH_3:Me_2S$

Crystalline: 90 %

Isofagomine

Scheme 11

Short synthesis of Isofagomine: 4 steps from the 5-bromo-5-deoxy-pentonolactone the isoiminosugar, as exemplified for the synthesis of isofagomine, now available from the bromolactone in four steps (*28*) (Scheme 11). Isofagomine inhibits β-glucosidase (almonds) with a Ki 0.11 μ M.

Using this simple strategy, we have been able to prepare several isoiminosugars in good yields (*27,28*). The inhibitory properties of the isoiminosugars synthesised are under investigation.

Highly Functionalised Cyclopentanes and Cyclohexanes: Synthesis of Carbasugars

A number of both naturally occurring and synthetic polyhydroxy aminocyclopentanes and –cyclohexanes have been found to be powerful inhibitors of glycosidases (*29*). The discovery of antibiotic and antitumor

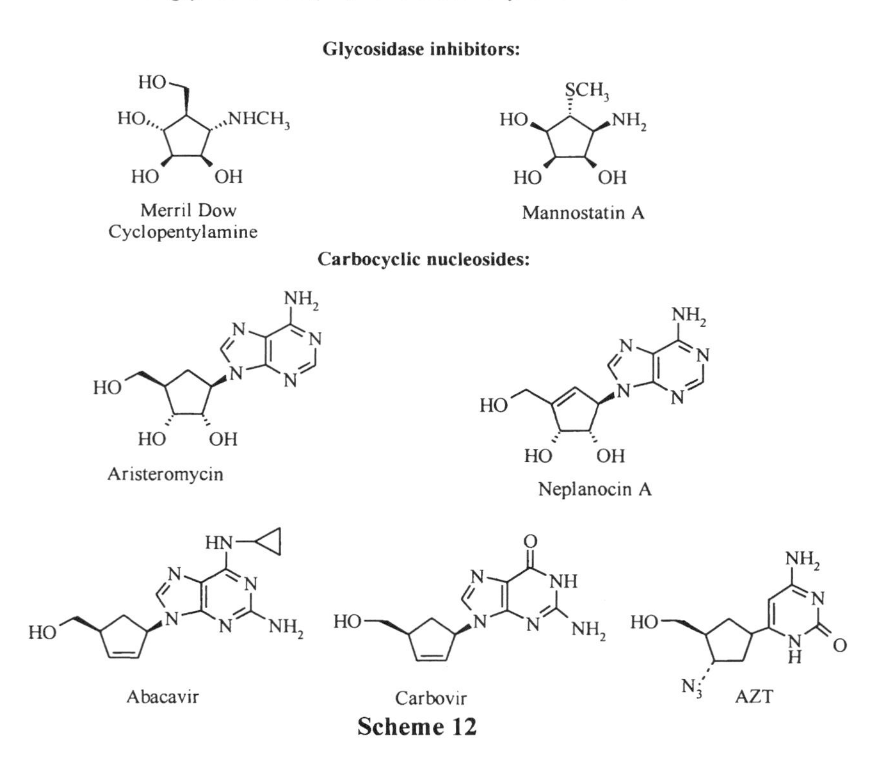

Scheme 12

Examples of known biologically active cyclopentane derivatives

activities of some naturally occurring carbonucleosides such as aristeromycin and neplanocin A has initiated the interest into the field of highly functionalised carbocyclic compounds and of carbocyclic nucleosides, and has resulted in an increasing number of synthetic procedures within this area (*30*). Carbovir, which undoubtly has been the most popular synthetic target among the carbocyclic nucleosides, was first synthesised in 1990 (*31*). It displays potent anti-HIV activity but is too toxic for clinical uses. The related carbanucleoside abacavir (*32*) is entering clinical use for dual therapy together with AZT for the treatment of HIV (*33*) (Scheme 12).

A number of different approaches for conversion of carbohydrate derivatives into functionalised carbocyclic compounds have been reviewed (*34*). Our approach towards the synthesis of this type of compounds has been the radical initiated carbocyclisation of ω-bromo-ω-deoxy-α,β-unsaturated aldonolactones.

The 2,3-unsaturated lactones are easily obtained from α,ω-dibromoaldonolactones, either by a reductive elimination, or by a base catalysed elimination of acetic acid yielding C-2 substituted 2,3-unsaturated lactones. Thus, when 2,7-dibromo-2,7-dideoxy-D-*glycero*-D-*ido*-heptono-1,4-lactone (**1**) was treated with sodium sulfite in methanol (*35*) the unsaturated lactone**2a** was formed in good yield (*36*), whereas pyridine or triethylamine eliminates a 3-*O*-actyl group from **1b** or **1c** to give the 2-*O*- or 2-*N*-substituted unsaturated lactones **2b** or **2c**, respectively (*37*) (Scheme 13). The protecting groups in the side chain must be alkyl rather than acyl groups, since the base will cause a further elimination of the latter. The lactones **1b** and **1c** are protected at C-5 and C-6 with an isopropylidene group.

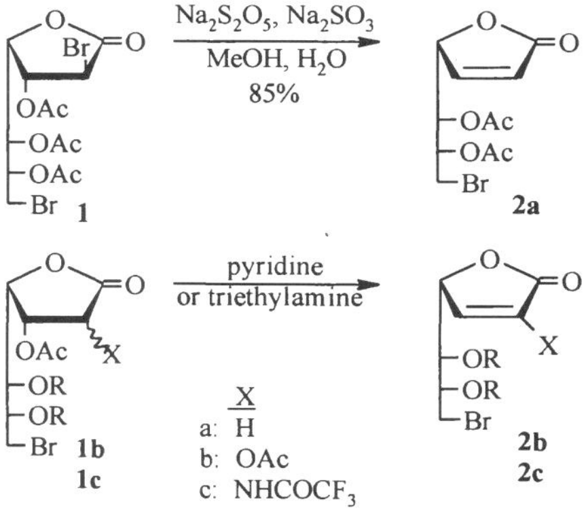

$Na_2S_2O_5$, Na_2SO_3; MeOH, H_2O; 85%

pyridine or triethylamine

X — a: H; b: OAc; c: $NHCOCF_3$

Scheme 13

Preparation of 2,3-unsaturated-aldono-1,4-lactones

Radical induced carbocyclisation of the unsaturated lactones **2a**, **2b** or **2c** in ethyl acetate using tributyltin hydride and a radical initiator gave in all cases stereospecifically *cis*-fused bicyclic cyclopentane lactones **3** in quantitative yields (Scheme 14). Thus, 7-bromo-2,3,7-trideoxy-D-*arabino*-hept-2-enono-1,4-lactone (**2a**) gave the bicyclic cyclopentane lactone **3a** in a quantitative yield as a single stereoisomer. Deacetylation gave the corresponding dihydroxy compound **4a** in a crystalline state (80% overall yield) (Scheme 14). Similar results were obtained from other unsaturated aldonolactones (*36*). When the carbocyclisation was performed using an unsaturated lactone having a substituent at C-2 (**2b** or **2c**) the bicyclic cyclopentane lactones **3b** or **3c** were formed with high stereoselectivity (>95%) with regard to the configuration at C-4 (Scheme 14). The success of generating stereospecifically the second new chiral center might be attributed to the roof-shaped conformation of the bicyclic system, thus protecting the endo phase from being trapped by the tinhydride in the final step (*36a,37a*). *The configurations of the two new chiral centers formed were thus both dictated by the configuration at C-4 in the starting lactone **2**: the protons at C-1, C-4 and C-5 in the products **3** were all* cis.

The lactone function in the bicyclic cyclopentane-lactones can be reduced to a hydroxymethyl group using sodium or calcium borohydride, whereby carbahexo- and –pentofuranoses were obtained (*36,37*) (Scheme 14).

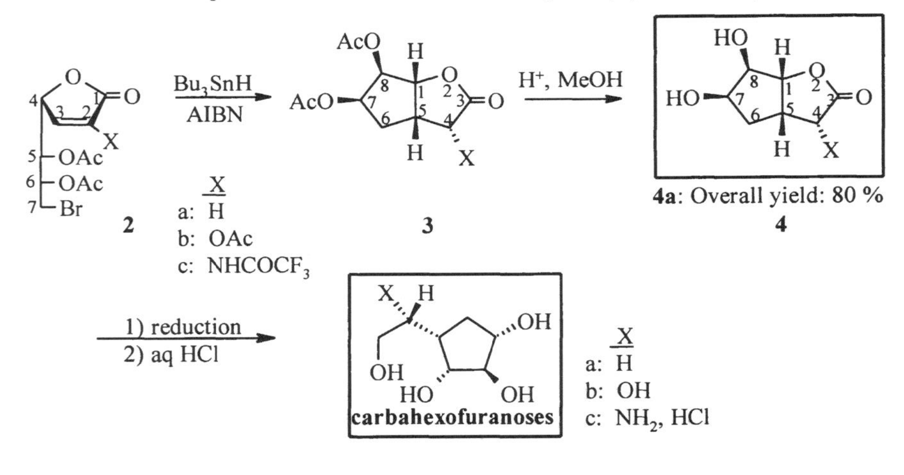

Scheme 14

Stereospecific carbocyclisation: formation of a bicyclic cyclopentane-lactone. Synthesis of carbahexofuranoses

When radical induced carbocyclisations of 8-bromo-2,3-unsaturated octonolactones *i.e.* **5** or **8** (Scheme 15) were performed, *cis*-fused cyclohexane-lactones **6** or **9** were formed with stereospecific generation of one or two new chiral centers (*38*) (Scheme 15). Again the configuration of the new chiral centers were determined by the configuration at C-4 in the unsaturated lactone. Deprotection followed by reduction of the lactone function gave the carbocyclic analogues of heptopyranoses, **7** or **10**.

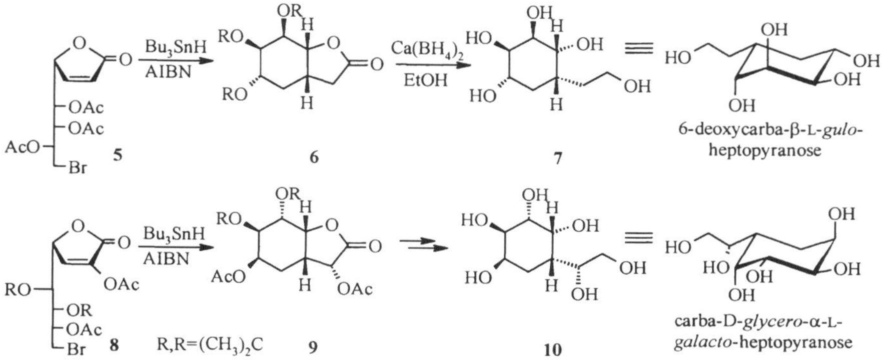

Bu_3SnH, AIBN; $Ca(BH_4)_2$, EtOH

R,R=$(CH_3)_2C$

6-deoxycarba-β-L-*gulo*-heptopyranose

carba-D-*glycero*-α-L-*galacto*-heptopyranose

Scheme 15

Stereospecific carbocyclisation: Synthesis of carbaheptopyranoses

As a conclusion, it should be emphasised that 2,3-unsaturated aldonolactones, in which a radical can be generated at the primary position, are useful starting materials for the stereospecific synthesis of highly functionalised, optically pure carbocyclic compounds.

Stereoselective Modification of the Bicyclic Cyclopentane-Lactone

The inexpensive, commercially available D-*glycero*-D-*gulo*-heptono-1,4-lactone ("glucoheptonolactone" since it is prepared from D-glucose) is the precursor for the 2,7-dibromo-2,7-dideoxy-D-*glycero*-D-*ido*-heptono-1,4-lactone (**1**) (Scheme 13) which, *via* the unsaturated lactone **2a** and radical induced carbocyclisation, could be transformed into the crystalline dihydroxy cyclopentane lactone **4a** in high yield (Scheme 14). This short and efficient synthesis of the *cis*-fused cyclopentane lactone provides easy access to a chiral synthon which might be modified stereospecifically for many purposes. Such successful transformations to valuable chiral molecules would thus only rely on the inexpensive "glucoheptonolactone". Our aim was now to study such transformations of **4a**: to change the configurations of the hydroxy groups, to introduce amino groups and to functionalise all carbon atoms within the cyclopentane ring. Subsequent reduction of the lactone ring would then result in chiral, highly functionalised cyclopentanes, which also might be viewed as carbasugar derivatives.

From the bicyclic diol **4a** we have prepared three new bicyclic chiral synthons: a bromohydrin **12**, an epoxide **14** and an allylic acetate **13** (Scheme 16). Thus, reaction of the diol with hydrogen bromide in acetic acid gave a *trans* bromo-acetate, **11**, with bromine introduced regio- and stereopecifically at C-7. Deacetylation to **12** followed by treatment with potassium carbonate in acetone gave the 7,8-*cis*-epoxide **14** (*39*). Treatment of **11** with an organic base (DBU) introduced a double bond between C-6 and C-7 to give the allylic acetate **13** (*40*). Hereby three new chiral synthons are accessible for further stereospecific modifications, taking advantage of the steric demanding roof-shaped bicyclic system. In the following only the preparation of nitrogen substituted carbocyclic derivatives, due to their possible biological activity, will be discussed, while other transformations have been performed as well (*39,40,41*).

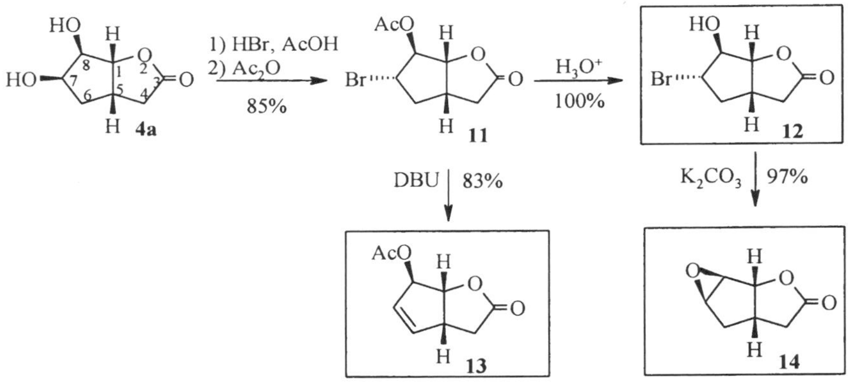

Scheme 16

*Preparation of 3 new chiral building blocks from **4a**.*

Synthesis of Amino/Hydroxy Substituted Cyclopentanes and a Carbocyclic Nucleoside

Nucleophilic substitution of the bromohydrine **12** as well as of the epoxide **14** was studied in order to introduce a nitrogen substituent at the carbocyclic ring. Azide substitution of **12** gave a C-7 azido derivative (*39*) which was also the case by opening of the epoxide **14** with the same nucleophile. Similarly, the epoxide was opened by ammonia at C-7 exclusively to give the *trans* amino alcohol **15** (*39*) (Scheme 17). Using a Ritter type reaction, in which acetonitrile acts as the nucleophile in the presence of an acid or of a Lewis acid, the nucleophile again opened the epoxide at C-7 to give the *trans* acetamido alcohol **16** (*39*). Accordingly, steric and not electronic requirements might be responsible for this regioselectivity in both a "SN2 and SN1 like" substitution reaction.

Reduction of the lactone ring of **15** yielded the aminocyclitol **17**, 1-amino-1,5-dideoxycarba-β-L-*xylo*-hexofuranose (Scheme 17).

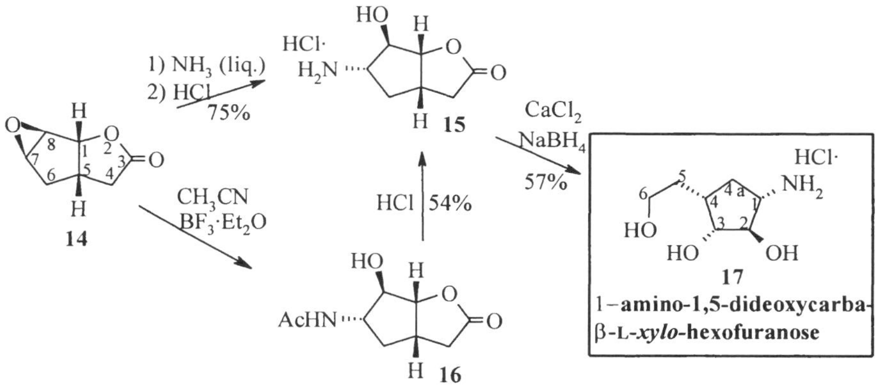

Scheme 17

Synthesis of amino/hydroxy substituted cyclopentanes: carbasugars

In order to functionalise the methylene carbon in the cyclopentane ring, we took advantage of the allylic acetate **13** (Scheme 16). Deacetylation gave the allylic alcohol **18** which by epoxidation of the double bond gave the 6,7-epoxide **19**, with the epoxy group in a *cis* relation to the C-8 hydroxygroup (Scheme 18). Nucleophilic opening of the epoxide with sodium azide, introduced regiospecifically the azido group at C-7. Reduction of the lactone and of the azido group gave the fully substituted amino hydroxy cyclopentane **22** (*40*) (Scheme 18). The two amino cyclopentanols **17** and **22** can be used as precursors for the synthesis of carbanucleosides, by building the heterocyclic base from the amino group, according to literature methods.

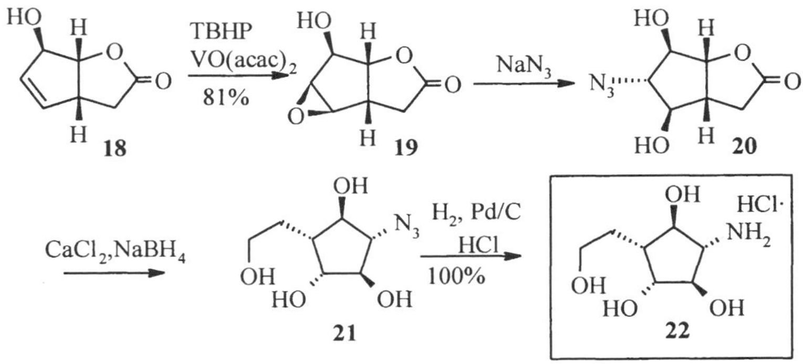

Scheme 18

Synthesis of densely functionalised, optically active cyclopentanes

Further studies on the use of the unsaturated compounds **13** or **18** as chiral building blocks have been published (*41*).

As pointed out above there is an interest in having access to new unsaturated carbanucleosides due to their biological activities, and we have investigated the possibility for using the bicyclic allylic alcohol **18** as a starting material in that context. Thus, substitution of the allylic hydroxy group with 6-chloropurine ina Mitsonubo reaction gave in a non-optimized reaction 20% of the nucleoside analogue (**23**). Reduction of the lactone moity gave **24** which by reaction with ammonia gave the unsaturated nucleoside (**25**) (Scheme 19 (*41*).

The similarity of the carbonucleoside **25** with the newly synthesised epinor-BCA (*42*), with promising activity against HIV, might indicate similar valuable acitvity for **25**. It remains, however, to be evaluated by testing.

Conclusion

Carbohydrate lactones have proven to be versatile starting materials for a range of complex target molecules. In this review the emphasis has been made to highlight the superior strategies from aldonolactones, when planning the synthesis of iminosugars and carbasugars *i.e.* highly functionalised pyrrolidine/piperidine or cyclopentane/hexane derivatives in an optically pure state. The importance of the optical purity of such biologically active compounds can not be overestimated, and thus the "chiral pool" should be considered in comparison with alternative methods by asymmetric synthesis.

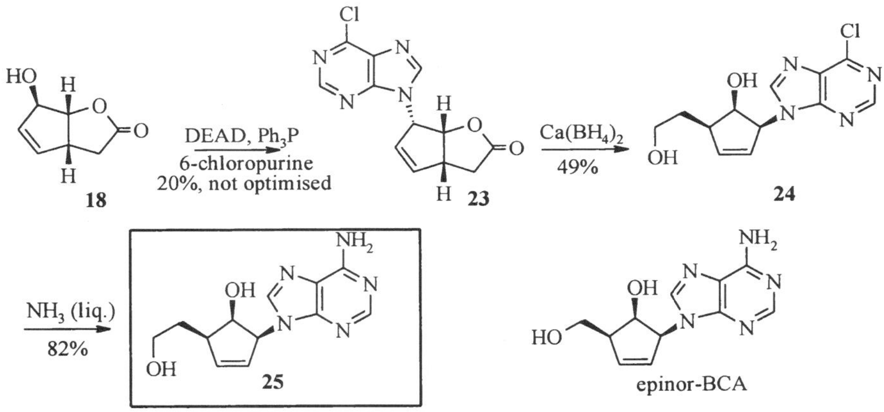

HO H O O H 18

DEAD, Ph_3P
6-chloropurine
20%, not optimised

Cl N N N N H O O H 23

$Ca(BH_4)_2$
49%

Cl N N OH N N OH 24

NH_3 (liq.)
82%

NH_2 N N OH N N OH 25

NH_2 N N OH N N HO epinor-BCA

Scheme 19

Synthesis of a carbanucleoside

References

1. a) Rassu, G.; Zanardi, F.; Battistini, L.; Casiraghi, G. *Synlett* **1999,** 1333; b) Lundt, I. *Top Curr Chem* **1997**, *187*, 117; c) de Lederkremer, R. M.; Varella, O. *Adv Carbohydr Chem Biochem* **1994**, *50*, 125; d) Fleet, G. W. J. Sugar Lactones as Starting Materials. In: *Antibiotics and Antiviral Compounds: Chemical Synthesis and Modifications*; Editors: Krohn K, Kirst HA, Maas H, **1993** VCH, Weinheim p 333.
2. Lundt, I.; Madsen, R. Iminosugars as powerful glycosidase inhibitors – synthetic approaches from aldonolactones. In: *Iminosugars as glycosidase inhibitors – nojirimycin and beyond*; Editor: Stütz , A. E. **1999**, Wiley-VCH, Weinheim pp 93-111.
3. Lundt, I. Reaction on C: Oxidation, reudction and deoxygenation at the anomeric position. In: Glycoscience: Chemistry and chemical biology; Editors: Fraser-Reid B, Tatsuta K, Thiem J. **2000**, Springer, Heidelberg
4. Cyörgydeák, Z.; Pelyvás, I.F. Monosaccharide sugars– chemical synthesis by chain elongation, degradation and epimerisation. **1998**, Academic Press, San Diego, p 18.
5. Cyörgydeák, Z.; Pelyvás, I. F. Monosaccharide sugars – chemical synthesis by chain elongation, degradation and epimerisation. **1998**, Academic Press, San Diego, p 450.
6. a) Isbell, H. S.; Frush, H. L. *Carbohydr Res.* **1979**, *72*, 30; b) Czarnocki, Z.; Mieczkowski, I. B.; Ziólkowski, M, *Tetrahedron: Asymmetry* **1996**, *7*, 2711.
7. For an overview of the available bromodeoxylactones by this method, see ref 1b
8. a) Lundt, I.; Frank, H. *Tetrahedron* **1994***, 50*: 13285; b) Kold, H.; Lundt, I.; Pedersen, C. *Acta Chem Scand* **1994**, *48*, 675.
9. Bouchez, V.; Stasik, I.; Beaupère, D.; Uzan, R. *Carbohydr Res* **1997**, *300*, 139
10. Lundt, I.; Madsen, R. *Synthesis* **1992**, 1129.
11. a) Lundt, I.; Madsen, R. *Synthesis* **1993**, 714 and 720; b) *ibid.* **1995**, 787.
12. Lundt, I.; Madsen, R.; Al Daher, S.; Winchester, B. *Tetrahedron* **1994**, *50*, 7513.
13. Godskesen, M.; Lundt, I.; Madsen, R.; Winchester, B. *Bioorg. Med. Chem.* **1996**, *4*, 1857.
14. Andersen, S. M.; Ekhart, E.; Lundt, I.; Stütz, A. E. *Carbohydr. Res.* **2000**, *326*, 22.
15. a) Withers, S. G.; Namchuuk, M.; Most, R. In: *Iminosugars as glycosidase inhibitors – nojirimycin and beyond*; Editor: Stütz, A.E., Wiley-VCH, Weinheim **1999**, 188; b) Sinnott, M.L. *Chem. Rev.* **1990**, *90*, 1171.

16. *Iminosugars as glycosidase inhibitors – nojirimycin and beyond;* Editor: Stütz, A.E Wiley-VCH, Weinheim, **1999**.

17. Bernotas, R. C.; Papandreou, G.; Urbach, J.; Ganem, B. *Tetrahedron Lett.* **1990**, *31*, 3393.
18. Legler, G.; Stütz, A. E.; Immich, H. *Carbohydr. Res.***1995**, *272*, 17.
19. Winchester, B.; Barker, C.; Baines, S.; Jacob, G. S.; Namgoon, S. K.; Fleet, G. *Biochem. J.* **1990**, *265*, 277.
20. a) Moris-Varas, F.; Qian, X.-H.; Wong, C.-H. *J. Am. Chem. Soc.*, **1996**, 7647; b) Quian, X-H..; Moris-Varas, F.; Fitzgerald, M. C. *Bioorg. Med. Chem.*, **1996**, 2055; c) Le Merrer, Y.; Depezay, J.-C.; Dosbaa, I.; Geoffroy, S.; Foglietty, M. J. *Bioorg. Med. Chem.* **1997**, *5*, 519.
21. Fleet, G. W. J.; Shaw, A. N.; Evans, S. V.; Fellows, L. E. *J. Chem. Soc., Chem. Commun.*, **1985**, 841.
22. Limberg, G.; Lundt, I.; Zavilla, J. *Synthesis* **1999**, 178.
23. Legler, G. *Adv. Carbohydr. Chem. Biochem.* **1990**, *48*, 319.
24. Umezawa, H.; Aoyagi, T.; Komiyama, T.; Morishima,H.; Hamada, T.; Takeuchi, T. *J. Antibibot.*, **1974**, *12*, 963.
25. a) Jespersen, T. M.; Dong, W.; Sierks, M. R.; Skrydstrup, T.; Lundt, I.; Bols, M. *Angew, Chem. Int. Ed. Engl.*, **1994**, *33*, 1778; b) Jespersen, T. M.; Bols, M.; Sierks, M. R.; Skrydstrup T. *Tetrahedron*, **1994**, *50*, 13449.
26. Lundt, I.; Madsen, R. Isofagomine and Beyond. In: *Iminosugars as glycosidase inhibitors – nojirimycin and beyond*; Editor: Stütz, A. E. **1999**, Wiley-VCH, Weinheim, pp 112-124.
27. Hyldtoft, L.; Godskesen, M.; Lundt, I. **2000** *to be published.*
28. Wagner, S.; Lundt, I. *to be published.*
29. Bereubar, A.; Grandjean, C.; Siriwardena, A. *Chem. Rev.* **1999**, *99*, 779.
30. a) Crimmins, M. T. *Tetrahedron* **1998**, *54*, 9229; b) Agrofolio, L.; Suhas, E.; Farese, A.; Condom, R.; Challand, S. R.; Earl, R. A.; Guedj *Tetrahedron* **1994**, *50*, 10611; c) Borthwick, A. D.; Biggadike, K. *Tetrahedron* **1992**, *48*, 571.
31. Vince, R.; Hua, M. *J. Med. Chem.* **1990**, *33*, 17.
32. a) Daluge, S. M. U.S. Patent 5.034.394 **1991**; b) Daluge, S. M.; Good, S. S.; Faletto, M. B.; Miller, W. H.; St. Clair, M. H.; Boone, L. R.; Tisdale, M.; Parry, N. R.; Reardon, J. E.; Dornsife, R. E.; Averett, D. R.; Krenitsky, T. A. *Antimicrob. Agents. Chemother.* **1997**, *14,* 1082.
33. a) Foster, R. H.; Faulds, D. *Drugs* **1998**, *55*, 729; b) Hamilton, C. J.; Roberts, S. M. *J. Chem. Soc., Perkin Trans. 1* **1999**, 1051.
34. Ferrier, R. J.; Middelton, S. *Chem Rev.* **1993**, *93*, 2779.
35. Vekemans, J. A. J. N.; Franken, G. A. M.; Dapperns, C. W. N.; Godefroi, E. F.; Chittenden, G. J. F. *J. Org. Chem.* **1988**, *53*, 627.

36. a) Horneman, A.M.; Lundt, I. *Synlett* **1995**, 918; b) ibid. *Tetrahedron* **1997** *53*, 6879.
37. a) Horneman, A.M.; Lundt, I. *J. Org. Chem.* **1998**, *63*, 1919; b) ibid. *Synthesis* **1999**, 317.
38. Lundt, I.; Wagner, S. H. *J. Chem. Soc., Perkin Trans 1* **2001**, 780.
39. Johansen, S. K.; Kornø, H. T.; Lundt, I. *Synthesis* **1999**, 171.
40. Johansen, S. K.; Lundt, I. *J. Chem. Soc., Perkin Trans 1* **1999**, 3615.
41. Johansen, S. K.; Lundt, I. *J. Org. Chem.* **2001**, *66*, 1129.
42. a) Katagiri, N.; Nomura, M.; Sato, H.; Tameda, C.; Kurimoto, A.; Arai, S.; Toyota, A.; Kaneko, C. *Nucleic Acids Symp. Series* **1991** *25*, 5; b) Katagiri, N.; Nomura, M.; Sato, H.; Kaneko, C.; Yusa, K.; Tsuruo, T. *J. Med. Chem.* **1992**, *35*, 1882.

Chapter 8

Rigid Polycycles and Peptidomimetics from Carbohydrate Synthons

Francesco Peri[1], Laura Cipolla[1], Barbara La Ferla[2], Eleonora Forni[1], Enrico Caneva[2], Luca De Gioia[1], and Francesco Nicotra[1,*]

[1]Department of Biotechnology and Biosciences, University of Milano-Bicocca, Piazza delle Scienze 2, I 20126 Milano, Italy
[2]Department of Organic and Industrial Chemistry, University of Milano, Via Venezian 21, 20133 Milano, Italy

The synthesis of conformationally constrained bicyclic and tricyclic compounds, obtained via 5-*Exo-Trig* iodocyclisation of polybenzylated sugars with an allylic substituent at the anomeric position, is described. The conformation of these molecules has been studied by n.O.e. experiments and Molecular Dynamics calculations. The introduction of an amino and a carboxylic group resulted in the formation of conformationally constrained bicyclic glyco-aminoacids that mimic protein turn conformation.

The design of new drugs with improved activity is frequently based on the introduction of conformational constraints in order to freeze bioactive conformations of the molecules. For this purpose, a variety of conformationally constrained scaffolds, that can be functionalized with different pharmacophores, have been developed. In this context, carbohydrates appear to be ideal substrates due to their conformational rigidity and offer the possibility of different functionalization of their hydroxyl groups. (*1*) On the other hand, the introduction of an amino and a carboxylic group into the rigid structure of the carbohydrate, allows to obtain conformationally constrained peptidomimetics, once the rigid glyco-aminoacid is incorporated into a peptide sequence. Pursuing this aim, the synthesis of different glyco-aminoacids has been performed, and the conformational analysis shows that these molecules can induce linear, β-turn or γ-turn conformation in peptides. (*2*) Azasugars, in which the amino group is already present in the cyclic skeleton, have also been exploited as analogues of aminoacids, and in particular of L-proline, provided that a carboxylic moiety is present in the structure. (*3*) The structural variety of glyco-aminoacids obtained so far is however limited to the well established pyranosidic and furanosidic forms bearing the amino and the carboxylic groups in different positions of the cycle.(*2,4*) We investigated the possibility of forming bicyclic structures which modify and enhance the rigidity of the sugar conformation, and allow to locate the amino and carboxylic groups in new relative orientations.

Bicyclic structures from carbohydrates

Sugars constrained in their energetically preferred or in a different conformation, have attracted the interest of synthetic chemists since a long time. An emblematic example is the 1,6-anhydro-D-glucose **1** (Figure 1), a commercially available compound in which the preferred $^4C_1$ conformation of the D-glucose is reversed into a $^1C_4$ conformation. More recently, rigid bicyclic carbohydrate-based structures have been developed by Fleet for the production of target oriented combinatorial libraries. (*5*) In the field of azasugars, it has been postulated that in castanospermine **2**, the presence of the five-membered ring is responsible for the high activity of the molecule as inhibitor of α-glucosidase, because it obliges the hydroxyl group corresponding to the C-6 hydroxyl group of D-glucose in a perpendicular orientation with respect to the plane of the piperidine cycle. (*6*)

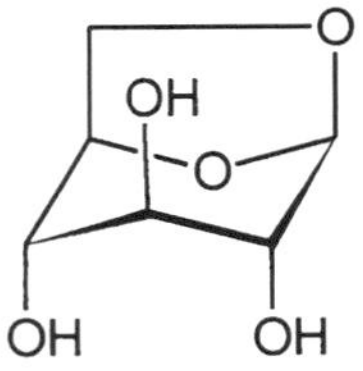

OH O OH O OH

1

OH HO N HO OH

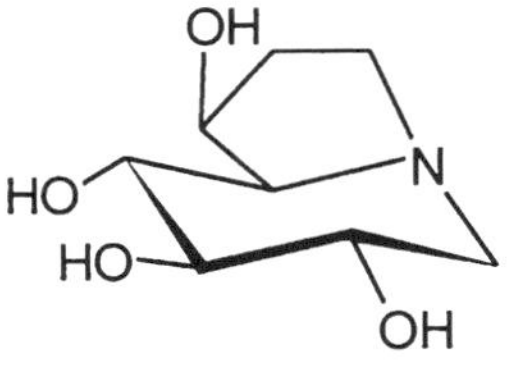

2

Figure 1

Furthermore, spiro-heterocycles derived from sugars are attractive targets, and some representative molecules of this class, such as hydantocidin **3** (Figure 2), are interesting bioactive molecules. (*7*)

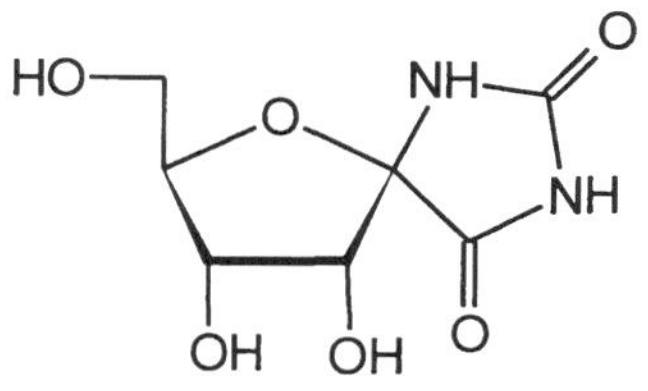

3

Figure 2: Hydantocidin

In 1987, studying the possibility of using different electrophiles in Sinaÿ's C-glucosydation procedure (Scheme 1) based on the Wittig-mercuriocyclization approach, (*8*) we observed that the use of iodine in oxolane-water at pH 4 resulted in the formation of the furanosidic structure **6**. (*9*)

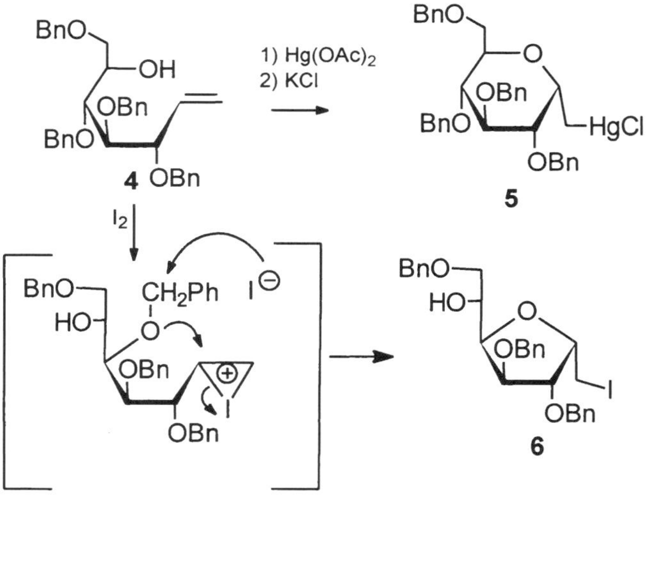

Scheme 1

This result clearly indicates that the *5-Exo-Trig* cyclization is strongly favoured, the iodonium ion being easily attached by the γ-benzyloxy group despite the presence of a free hydroxyl group in δ position. The process results in a regioselective debenzylation, with formation of a cyclic iodoether.

We took advantage of this observation in order to effect selective debenzylations, once an allylic appendage is attached to the sugar. For example, when the polybenzylated allyl-C-glucopyranoside **7** (*10*) is treated with iodine in THF at 0 °C, the bicyclic compound **8** is obtained (85:15 in favor to the 2'*R* isomer), which upon treatment with zinc and acetic acid undergoes a reductive elimination affording the C-glucoside **9** selectively deprotected at C-2. (*11*)

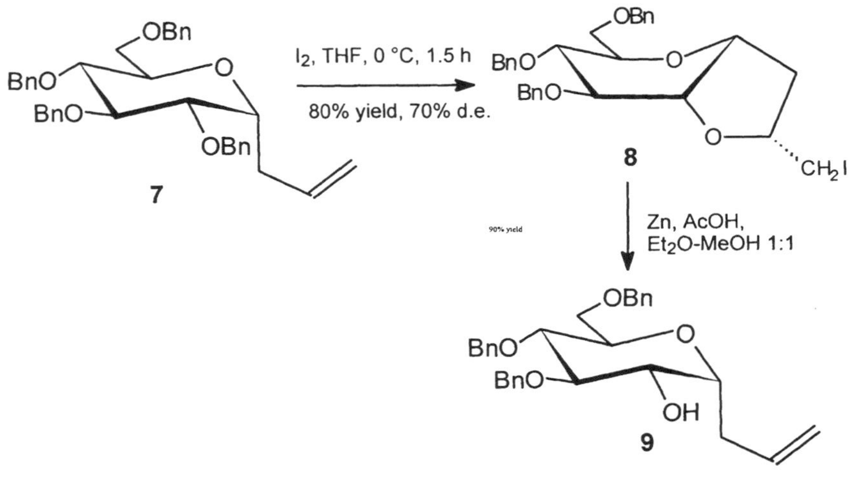

OBn
BnO
BnO
O
OBn
7
I_2, THF, 0 °C, 1.5 h
80% yield, 70% d.e.
OBn
BnO
BnO
O
O
8
CH_2I
Zn, AcOH,
Et_2O-MeOH 1:1
90% yield
OBn
BnO
BnO
O
OH
9

Scheme 2

This selective deprotection has been exploited in order to obtain C-glycosides of aminosugars, the synthesis of which is complicate when starting from the corresponding aminosugar. (*11*)

The iododebenzylation approach seams to be quite promising in order to obtain conformationally constrained bicyclic structures. With this idea in our mind we have inserted an allylic appendage at the anomeric center of different sugars. This can be effected by Lewis acid catalyzed reaction of the sugar with allyltrimethylsilane obtaining a C-glycopyranoside with an α anomeric configuration, (*10*) or by reaction of the glyconolactone with allylmagnesium bromide, followed by Lewis acid catalyzed reduction of the obtained lactol with triethylsilane, affording the β-C-glycopyranoside. The two processes are complementary from the stereochemical point of view, and we observed that the iodocyclization-debenzylation also occurs on the β-C-glucopyranoside **10**, where the allylic appendage is equatorially oriented. In this case the reaction is much slower, requiring 5 hrs at rt. for the stereoselective formation of the bicyclic iodide **11**. (*11b*)

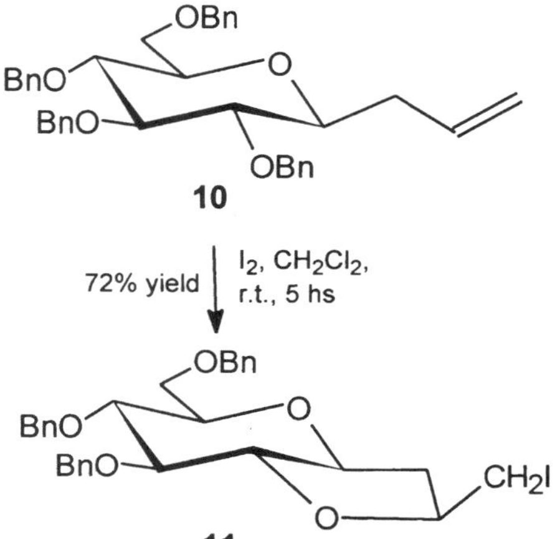

10

72% yield | I_2, CH_2Cl_2, r.t., 5 hs

11

Scheme 3

By the same reaction sequence, D-arabinofuranose was transformed into a bicyclic scaffold (Scheme 4). Commercially available 2,3,5-tri-O-benzyl-D-arabinofuranose was converted into the anomeric acetate **12**, which was allylated by treatment with allyltrimethylsilane in the presence of a catalytic amount of boron trifluoride etherate. The reaction, effected at rt. in acetonitrile, afforded **13** as a 1:1 mixture of α and β diastereoismers (72% yield). The iodocyclization was effected with iodine in dichloromethane at 0 °C. Only the β diastereoisomer reacted in 12 hrs, affording the bicyclic iodoether **14** as a mixture of diastereoisomers (49% yield). The diastereoisomer with 2'-(S) configuration, was formed in about 20% d.e. **14-R** and **14-S** were separated by flash chromatography and their absolute configuration at the 2' carbon was assigned by a NOESY analysis. In compound **14-R** a consistent NOESY crosspeak between H-2' and H-3 protons is diagnostic for the (R) configuration. The opposite (S) configuration in compound **14-S** is indicated by a strong n.O.e. between H-2' and H-2.

A spiro bicyclic structure was obtained by treatment of the α-allyl-C-fructofuranoside **15** (*12*) with iodine (Scheme 5). The reaction afforded the spiro derivative **16**, which once more resulted from the 5-*Exo-Trig* cyclization, involving a benzyloxy group and consequent debenzylation.

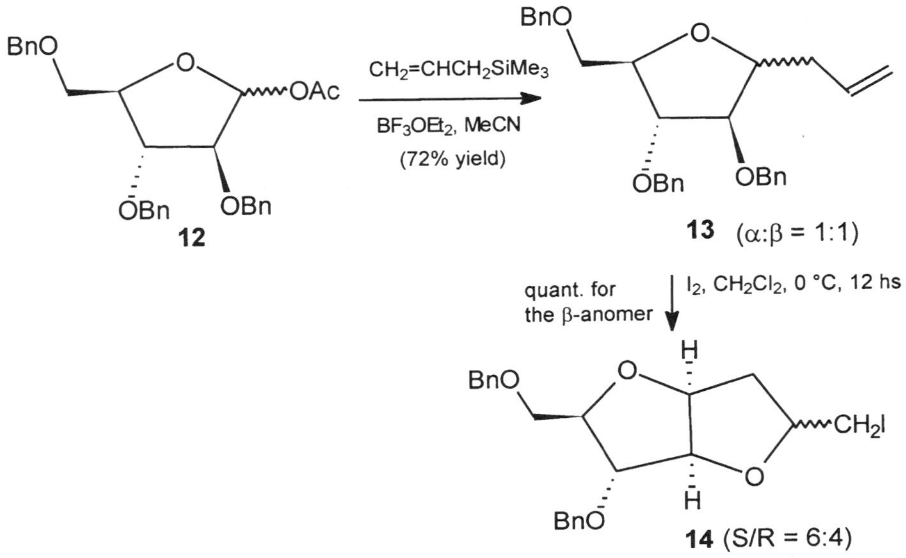

BnO OAc OBn OBn **12**

$CH_2{=}CHCH_2SiMe_3$

BF_3OEt_2, MeCN

(72% yield)

BnO OBn OBn **13** (α:β = 1:1)

quant. for the β-anomer

I_2, CH_2Cl_2, 0 °C, 12 hs

BnO H CH_2I H BnO **14** (S/R = 6:4)

Scheme 4

BnO OBn BnO OBn **15**

I_2, THF

rt, 0.5 h

98% yield

BnO BnO CH_2I OBn **16** (33% d.e.)

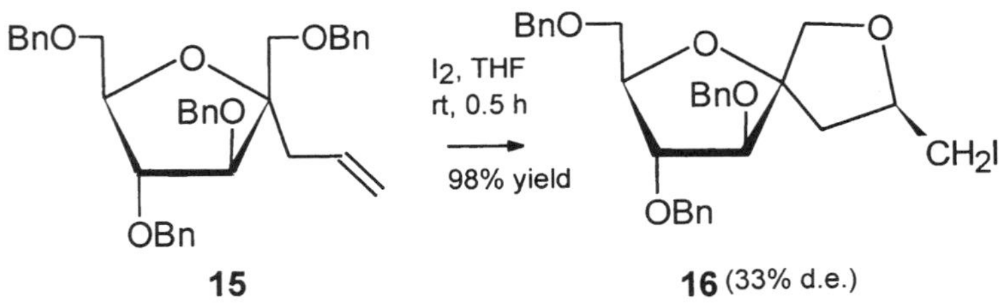

Scheme 5

The reaction stereoselectively affords the 2'-*R* isomer as the major product (d.e. 33%). The absolute configuration of the 2' carbon was determined by NOESY analysis, in particular consistent n.O.e.s between H-1 and H-1'b and between H-1'a and H-2' protons were observed. Treatment of **16** with zinc and acetic acid, afforded **17**, that was transformed in the C-fructoside **18** (Scheme 6). Compound **18** can be considered as an α- or a β-C-fructoside, in which both the hydroxymetyl "arms" can be further manipulated. An interesting example of this is the formation of the new spiro structure **20**, as reported in Scheme 6.

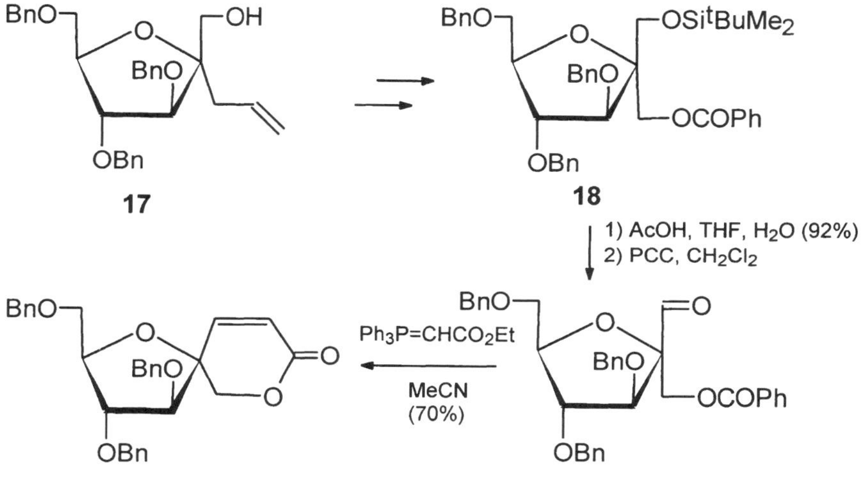

Scheme 6

Selective deprotection of the silyl ether by hydrolysis, oxidation of the free hydroxyl group to the aldehyde **19** and then reaction with carboethoxymethylenetriphenylphosphorane directly afforded the spiro compound **20**, which is an interesting bicyclic Michael acceptor.

Alternatively, compound **17** was oxidized to the aldehyde **21** (Scheme 7), according to the Swern procedure, then treated with vinylmagnesium bromide in THF at room temperature to afford compound **22**, in which the two double bonds can be exploited for further cyclizations. Compound **22** was obtained in a very high diastereomeric excess (97.5% determined by HPLC), reasonably due to a Cram-chelated intermediate in which magnesium coordinates to the carbonyl oxygen and the oxygen of the furanosidic cycle (Figure 3), the attack of the Grignard reagent occurring from the less hindered *re* face of the carbonyl group.

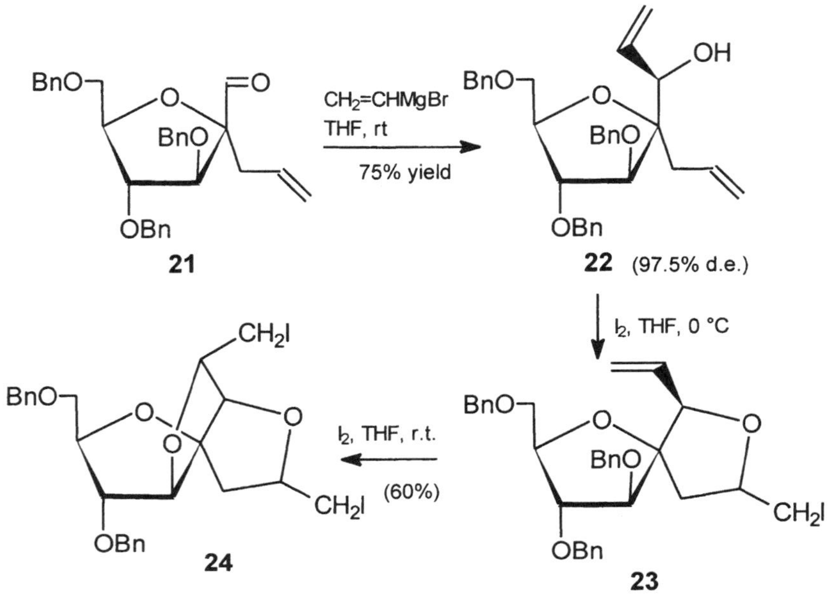

Scheme 7

Treatment of compound **22** with an excess of iodine in THF at 0 °C resulted in the formation of compound **23**, the free hydroxyl group being favored in the nucleophilic attack to the iodonium ion. It is worth of note that, despite the presence of two double bonds, at 0 °C the reaction is regioselective. Increasing the temperature to 20 °C, the second iodocyclization with debenzylation occurs, affording compound **24** in 60% overall yield.

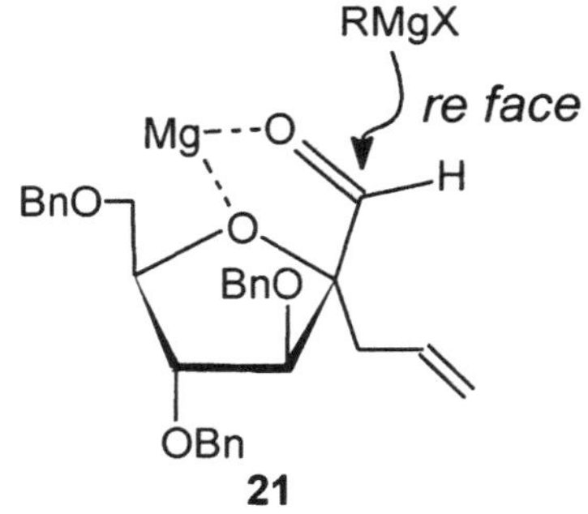

Figure 3

The iodocyclization reaction can also be extended to azasugars, provided that one can introduce an allylic appendage at the "anomeric" centre of this class of molecules. The synthesis of an allyl "C-glycoside" of nojirimycin has been reported by R. R. Schmidt[13] from glucose through the key intermediate 1-fluoro-1-deoxynojirimycin, using a complex multistep sequence. We have developed a new and efficient approach which allows the stereoselective synthesis of polybenzylated α-1-allyl-1-deoxynojirimycin in 6 steps and 36% overall yield, using tetrabenzylglucose as commercially available starting material. The synthetic strategy, reported in Scheme 8, requires first the introduction of the amino function and the allylic appendage and finally the cyclization to the desired piperidine derivative. From a stereochemical point of view, the allylation and the cyclization reactions are crucial to the effectiveness of the synthesis, and must be highly stereoselective

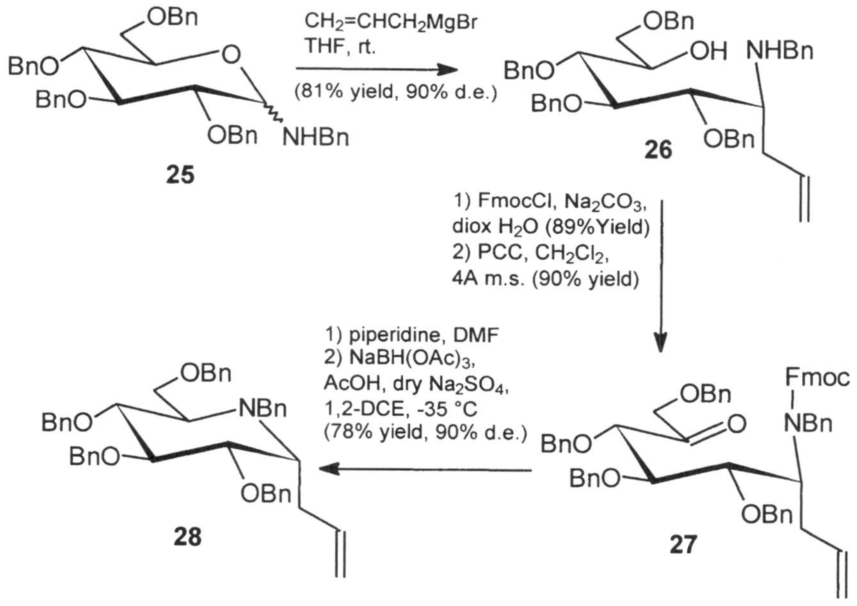

25 → 26: CH_2=$CHCH_2MgBr$, THF, rt. (81% yield, 90% d.e.)

26 → 27: 1) FmocCl, Na_2CO_3, diox H_2O (89%Yield) 2) PCC, CH_2Cl_2, 4A m.s. (90% yield)

27 → 28: 1) piperidine, DMF 2) $NaBH(OAc)_3$, AcOH, dry Na_2SO_4, 1,2-DCE, -35 °C (78% yield, 90% d.e.)

Scheme 8

The introduction of the amino group was performed by reaction of tetrabenzylglucose with benzylamine. The resulting glycosylamine**25** was then treated with allylmagnesium bromide in order to introduce stereoselectively (through a Cram-chelated intermediate), the allylic appendage. The cyclization of the open-chain intermediate **26** to the "allyl-α-C-glycoside" of nojirimycin **28**, was accomplished by PCC oxidation of the N-Fmoc protected derivative to ketone **27**, followed by deprotection and reductive amination with sodium triacetoxyboronhydride.

Preliminary results indicate that the polybenzylated azasugar **28** can be converted into the bicyclic derivative **29** (Figure 4), by treatment with N-iodosuccinimmide in THF at room temperature, taking advantage of the allylic appendage. In this reaction a single stereoisomer is obtained, the stereochemistry of which has still to be elucidated.

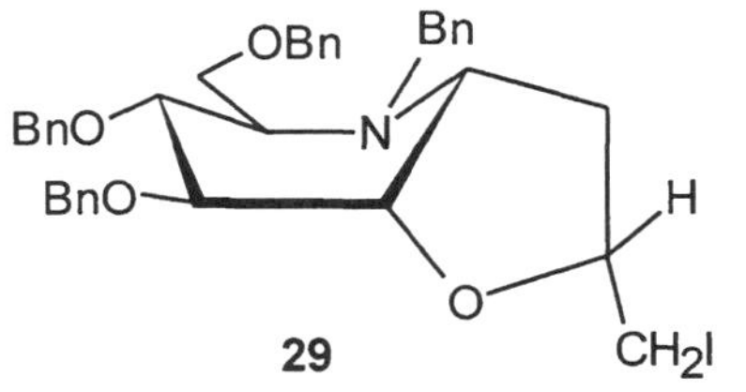

Figure 4

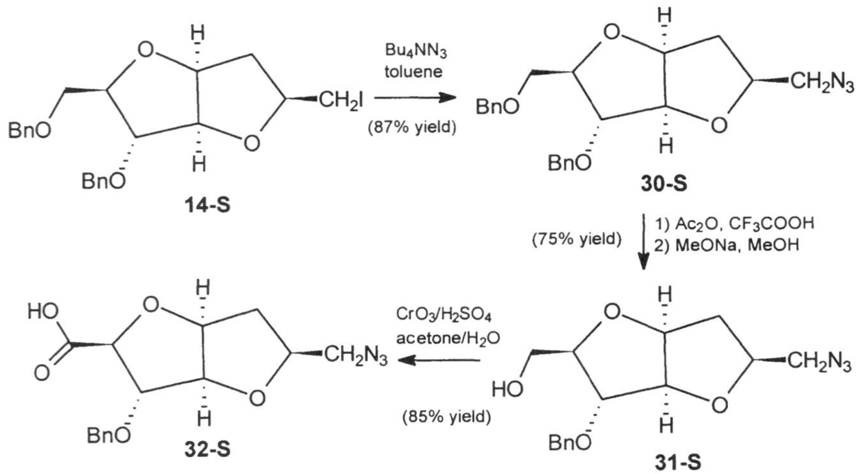

Scheme 9

Introduction of azido and carboxylic functions

As example of introduction of amino and a carboxylic functions in a rigid bicyclic glyco-structure, compound **14-S** was treated with tetrabutylammonium azide in toluene at 70 °C, affording the azide **30-S** in 87 % yield (Scheme 9).

In order to introduce the carboxylic function, azide **30** was regioselectively debenzylated at the primary hydroxyl group by controlled treatment with acetic anhydride and trifluoroacetic acid, (*14*) followed by saponification of the acetate at C-5. The product **31** was finally converted into the carboxylic acid **32** by Jones oxidation.

The same sequence of reactions was accomplished on the bicyclic iodide **16-R** (Scheme 10), the reaction of which with tetrabutylammonium azide in toluene at 60 °C afforded the azide **33-R** in 76% yield. Selective deprotection of the primary hydroxyl group of compound **33-R** and Jones oxidation, finally afforded compound **35-R** as masked glyco-aminoacid. The azido group was maintained in these compounds as protected equivalent of the amino group.

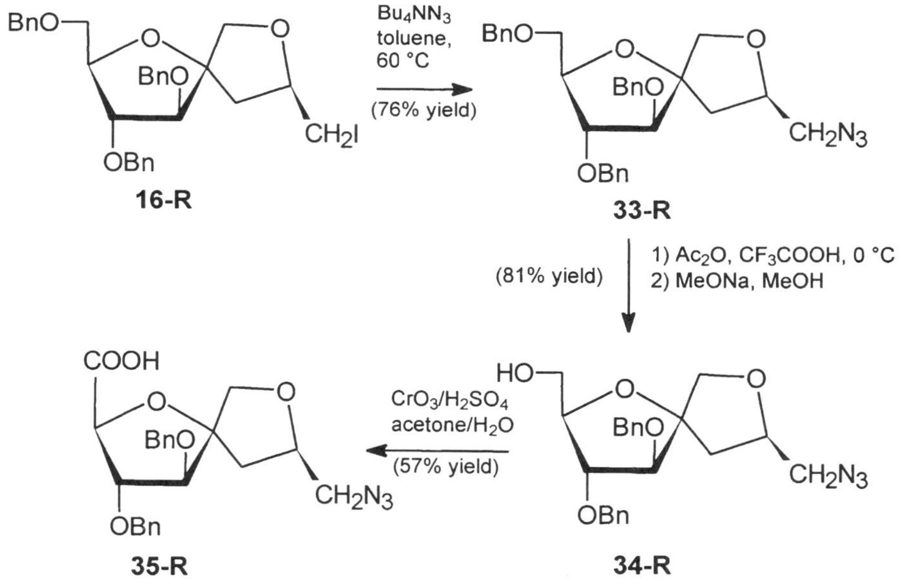

BnO O O BnO OBn CH_2I **16-R**

Bu_4NN_3 toluene, 60 °C (76% yield)

BnO O O BnO OBn CH_2N_3 **33-R**

1) Ac_2O, CF_3COOH, 0 °C 2) MeONa, MeOH (81% yield)

HO O O BnO OBn CH_2N_3 **34-R**

CrO_3/H_2SO_4 acetone/H_2O (57% yield)

COOH O O BnO OBn CH_2N_3 **35-R**

Scheme 10

Molecular dynamics and NMR structural studies

Molecular mechanics (MM) and dynamics (MD) are useful tools to investigate the conformational properties of organic molecules. (*15*) In particular, the combined use of MM and MD can be very effective in sampling the potential energy hypersurface (PES) when structurally constrained molecules are considered. In the present work, the PES has been described using the MM+ forcefield (*16*) and MM optimizations were followed by short MD runs (10 ps) carried out at different temperature (from 300 to 700 K) in order to sample the PES efficiently. Usually, due to the steric properties of the molecules investigated, no more than 10 MM/MD cycles were necessary to localize all the relevant energy minimum structures.

These energy minimum structures have been compared with the structures resulting from the interproton distances calculated through a collection of NOESY experiments using different mixing times (from 0.4 to 1.3 s). Extracting the maximum n.O.e. values for each nuclei couple from the curves indicating the variation of magnetization transfer in NOESY spectra, distances were calculated by comparison with geminal standard distance (1.8 Å). (*17*)

We investigated the conformation of the amino acids **36** and **37** corresponding to the azido acids **32** and **35** deriving respectively from D-arabinose and D-fructose (figure 5). We assumed that the conformation of the bicycle does not considerably change susbstituting the azido group with an amino group in position 3'.

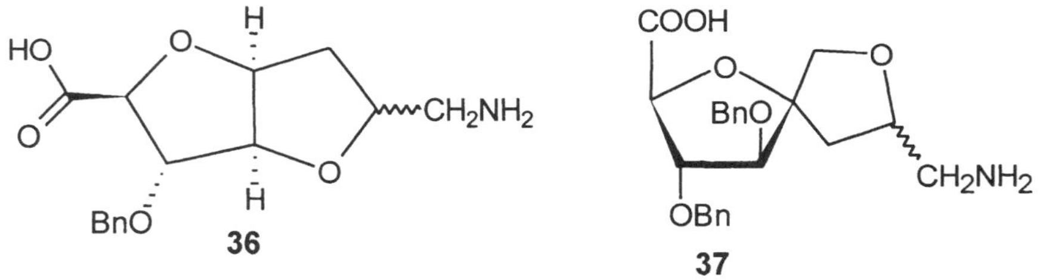

36 **37**

Figure 5

The computational investigation of **37-R** and **37-S** shows that, even if these molecules are characterized by steric constraints due to the spiro linkage of the two five-member rings, up to four slightly different conformers featuring similar energy were located by the MM/MD procedure. The comparison of the

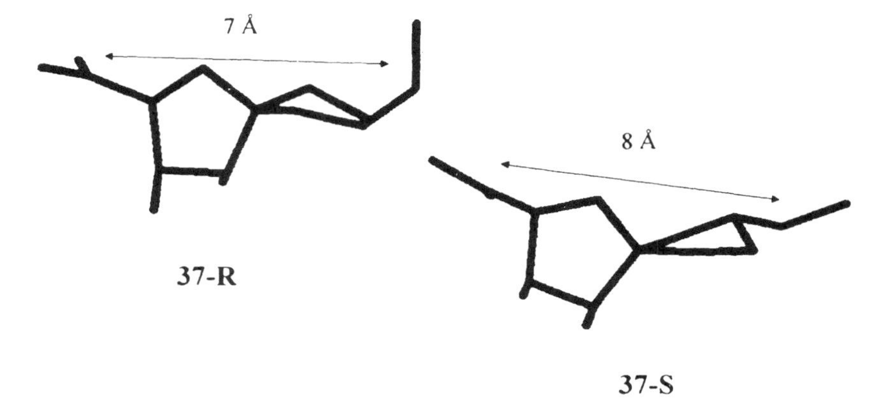

Figure 6

computational data with interproton distances obtained by NOESY experiments on the azides **35-R** and **35-S** allowed to refine the conformational search and converge on the structures schematically shown in Figure 6.

37-R and **37-S** are characterized by very similar, even if not identical, structural properties of the two 5-member rings. The distance between the C=O and the NH_2 groups is about 7 Å for **37-R** and 8 Å for **37-S**. Moreover, the orientation of the NH_2 group in **37-R** seems more suited to mimic protein turn mimetics.

The conformational analysis of **36-R** and **36-S** shows that these molecules are characterized by a high rigidity, due to the condensed five-membered rings. In fact, the MM/MD procedure always converged on the sameconformer family, as shown in Figure 6. In particular, **36-S** features a very small distance between the C=O and the NH_2 group (about 6 Å) and is characterized by a mutual orientation of these two groups that appears very favourable to mimic sharp turns. On the other hand, **36-R** is characterized by C=O/NH_2 distance of about 7 Å.

In conclusion, compound **36-S** is an excellent candidate as protein β-turn mimetic because of its conformational rigidity and the correct mutual orientation of the carboxy and amino groups. Compounds **37-R**, **37-S** and **36-R** can act as rigid spacers having a fixed distance of about 7-8 Å between the carboxylic and the amino groups. These compounds are good candidates to space the C- and N-terminal residues of protein Ω loops where, according to the definition of Leszczynsky and Rose, (*18*) the distance between segment termini is approx. 4-11 Å.

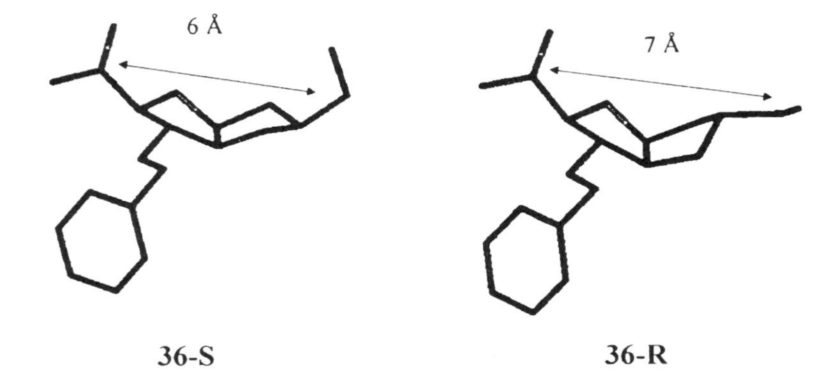

Figure 7

References

1. a) Wunberg, T.; Kallus, C.; Opatz, T.; Henke, S.; Schmidt, W.; Kunz, H. *Angew. Chem. Int. Ed.* **1998**, *37*, 2503; b) Hirschmann, R.; Yao, W.; Cascieri, M. A.; Strader, C. D.; Maechler, L.; Cichy-Knight, M. A.; Hynes, J.; van Rijn, R. D.; Spengeler, P. A., Smith III, A. B. *J. Med. Chem.* **1996**, *39*, 2441; c) Hirschmann, R.; Nicolau, K. C.; Pietronico, S.; Leay, E. M.; Salvino, J.; Arison, B.; Cichy M. A.; Spoors, P. G.; Shakespeare, W. C.; Sprengeler, P. A.; Hamley, P.; Smith III, A. B.; Reisine, T.; Raynor, K.; Maechler, L.; Donaldson, C.; Vale, W.; Freidinger, R. M., Cascieri, M. R.; Strader, C. D. *J. Am. Chem. Soc.*, **1993**, *115*, 12550; d) Nicolau, K. C.; Trujillo, J. I.; Chibale, K. *Tetrahedron*, **1997**, *53*, 8751; e) Hirschmann, R; Nicolau, K. C.; Pietranico, S.; Salvino, J.; Laehy, E. M., Splengeler , P. A.; Furst, G.; Smith III, A. B.; Strader, C. D., Cascieri, M. A.; Candelore, M. R. *J. Am. Chem. Soc.*, **1992**, *114*, 9217; f) Hirschmann, R.; Hynes Jr, J.; Cichy-Knight, M. A. van Rijn, R. D.; Sprengeler, P. A.; Spoors P. G.; Shakespeare, W. C.; Pietronico-Cole, S.; Barbosa, J.; Liu, J.; Yao, W.; Rohrer, S.; Smith II, A. B. *J. Med. Chem.* **1998**, *41*, 1382; g) Sofia, M. J.; Hunter, R.; Yau Chan, T.; Vaughan, A.; Dulina, R.; Wang, H.; Gange, D. *J. Org. Chem.* **1998**, *63*, 2802.
2. Von Roedern, E. G.; Lohof, E.; Hessler, G.; Hoffmann, M.; Kessler, H. *J. Am. Chem. Soc.* **1996**, *118*, 10156.
3. a) Bashyal, B. P., Chow, H-F.; Fleet, G. W. J. *Tetrahedron*, **1987**, *43*, 423; b) Bruce, I.; Fleet, G. W. J.; Cenci de Bello, I.; Winchester, B. *Tetrahedron*, **1992**, *48*, 10191; c) Park, K. H.; Yoon, Y. J.; Lee, S. G. *J. Chem. Soc. Perkin trans. 1*, **1994**, 2621.
4. a) Smith, M. D.; Long, D. D.; Marquess, D. G.; Claridge, T. D. W., Fleet G. W. J. *J. Chem. Soc. Chem. Comm*un. **1998**, 2039; b) Smith, M. D.; Claridge, T. D. W.; Tranter, G. E.; S.; Sansom, M. S. P.; Fleet, G. W. J. *J. Chem. Soc. Chem. Comm*un. **1998**, 2041; c) Suhara, Y.; Izumi, M.; Ichikawa, M.; Penno, M. B., Ichikawa, Y. *Tetrahedron Lett.*, **1997**, *38*, 7167; d) Nicolau, K. C.; Florte, H.; Egan, M. G.; Barth, T.; Estevez, V. A. *Tetrahedron Lett.* **1995**, *36*, 1775.
5. See for example a) Smelt, K. H.; Blérot, Y.; Biggadike, K.; Lynn, S.; Lane, A. L.; Watkin, D. J.; Fleet, G. *Tetrahedron Lett.*, **1999**, *40*, 3255; b) Smelt, K. H.; Arrison, A. J.; Biggadike, K.; Muller, M.; Prout, K.; Watkin, D. J.; Fleet, G. *Tetrahedron Lett.*, **1999**, *40*, 3259.
6. van der Broek, L. A. G. M.; Vermaas, D. J.; Heskamp, B. M.; van Beeckel C. A. A.; Tan, M. C. A. A.; Bolscher J. G. M.; Ploegh, H. L.; van Kemenade, F. J.; de Goede, R. E. Y.; Miedema, F., *Recl. Trav. Chim. Pays-Bas* **1993**, *112*, 82.

7. a) Osz, E.; Szilàgyi, L.; Somasàk, L.; Bényei, A. *Tetrahedron* **1999**, *55*, 2419; b) Hanessian, S.; Lu, P-P.; Sancéau, J-Y.; Chemla, P.; Gohda, K.; Fonne-Pfister, R.; Prade, L.; Cawan-Jacob, S. W. *Angew. Chem. Int. Ed.* **1999**, *38*, 3159.
8. Pougny, J-R.; Nassr, M.A.M.; Sinaÿ, P. *J. Chem. Soc. Chem. Commun.*, **1981**, 375.
9. Nicotra, F.; Panza, L.; Russo, G.; Toma, L. *Carbohydr. Res.* **1987**, 171, 49.
10. a) Lewis, M. D., Cha, J. K. and Kishi, Y. *J. Am. Chem. Soc.*, **1982**, *104*, 4976; b) Hosomi, A.; Sakata, Y.; Sakurai H. *Tetrahedron Lett.* **1984**, *25*, 2383.
11. a) Cipolla, L.; Lay, L.; Nicotra, F. *Carbohydr. Lett*, **1996**, *2*, 131; b) Cipolla, L.; Lay, L.; Nicotra, F. *J. Org. Chem.*, **1997**, *62*, 6678; c) Lay, L.; Cipolla, L.; La Ferla, B.; Peri, F.; Nicotra, F. *Eur. J. Org. Chem.*, **1999**, 3437.
12. Nicotra, F.; Panza, L.; Russo, G. *J. Org. Chem.*, **1987**, *52*, 5627.
13. Fuchss, T.; Streicher, H.; Schmidt, R. R. *Liebigs Ann. Recueil*, **1997**, 1315.
14. Eby, R.; Sondheimer, S. J.; Schuerch, C. *Carbohydr. Res.* **1979**, *73*, 273.
15. Leach, A.; Molecular modelling, principles and applications. Longman Ed. **1996**.
16. Allinger, N. L. *J. Am. Chem. Soc.* **1977**, *99*, 8127.
17. Mascagni, P.; Gibbons, W. A. *J. Chem. Soc. Perkin Trans. I* **1985**, 245
18. a) Leszczynsky, J. F.; Rose, G. D. *Science* **1986**, *234*, 849, b) Tramontano, A.; Chothia, C.; Lesk, A. *Proteins: structure, Function and Genetics* **1989**, 6, 382.

Chapter 9

Recent Progress in Total Synthesis and Development of Natural Products Using Carbohydrates

Kuniaki Tatsuta

Graduate School of Science and Engineering, Waseda University, Shinjuku, Toyko 169-8555, Japan

The total synthesis and development of a variety of natural products have been accomplished by using carbohydrates as chiral sources. The target molecules are nonsteroidal progesterone receptor ligands (PF1092A, B and C), antibiotics (pyralomicins), glyoxalase I inhibitor and its precursor KD16-U1, glycosidase inhibitors (valienamine and validamine), N-methyl-D-aspartate receptor antagonists (ES-242s) and one of big four antibiotics (tetracycline).

1. Introduction

All the structures of natural products are very beautiful and attractive. Then, I would like only to relate them to my favorite compounds, carbohydrates. In my opinion, carbohydrates are the language of chiral natural products; therefore, I have focused on the use of carbohydrates as chiral precursors in organic synthesis.

Herein, I would like to present our recent work in the total synthesis and development of medicinally useful natural products, which use carbohydrates as chiral sources to determine the absolute structure of the natural products and to clarify their structure - activity relationships (*1*).

2. The First Total Synthesis of Progesterone Receptor Ligands, PF1092A, B and C

The microbial metabolites (-)-PF1092A, B and C (**10**, **11** and **12**) were isolated as new nonsteroidal progesterone receptor ligands by the Meiji Seika group from the culture broth of *Penicillium oblatum*, and the absolute structures were finally determined by X-ray crystallographic analysis (*2*). Structurally, they belong to the complex eremophilane-type sesquiterpenes, with four contiguous *cis*-substituents on an octalone skeleton fused with a butenolide ring.

The first enantiospecific total synthesis of (-)-PF1092A, B and C (**10** - **12**) is based on the $SnCl_4$- promoted cyclization of an α-keto methyl sulfone and dimethyl acetal followed by a Stork annulation which gives the octalone core (*3*) (Scheme 1).

The critical step was the direct opening of the furanose ring (**4** to **5**) by silylation with simultaneous formation of the enol silyl ether, because the one-step opening of the furanose ring is generally difficult.

The synthesis was initiated with the stereoselective introduction of two methyl groups onto the tritylated butenolide **1** to give the dimethylated lactone **2** (67%) along with the C-2 epimer (13%). As this stereocenter will be lost in the Stork annulation (*vide post*), both epimers could be used in the total synthesis of **12**. Their structures were confirmed by the NOE enhancement in **2**. After detritylation, the resulting alcohol was transformed to the dimethyl acetal **3**. Reaction with the lithiated $MeSO_2Ph$ gave the lactol **4**, which was silylated to the open chain having the enol silyl ether **5** (91%). These reactions seem to depend on the readiness of the enol silyl ether formation.

After investigating various derivatives and Lewis acids, the desired aldol-type cyclization of **5** with β-elimination was realized by treatment with $SnCl_4$. Desulfurization with Al(Hg) with concomitant reduction of the olefin gave the cyclohexanone **6** (69%). These procedures feature general methods of entry into optically active cyclohexanes and cyclohexanols. The annulation of **6** was carried out according to Stork's procedure (*4*) by silylation to give the silyl enol ether, followed by successive treatment with a silylated methyl vinyl ketone and with MeONa to give the desired octalone **7** in 60% overall yield. The introduction of the ethyl methyl ketone moiety to C-2 in **6** was expected to occur with addition *trans* to the C-3 methyl group to afford the natural configurations in **7**. The NOE enhancement was clearly detected between two methyl signals to support the *cis*-dimethyl structure. Compound **7** was converted into the Zn enolate and reacted

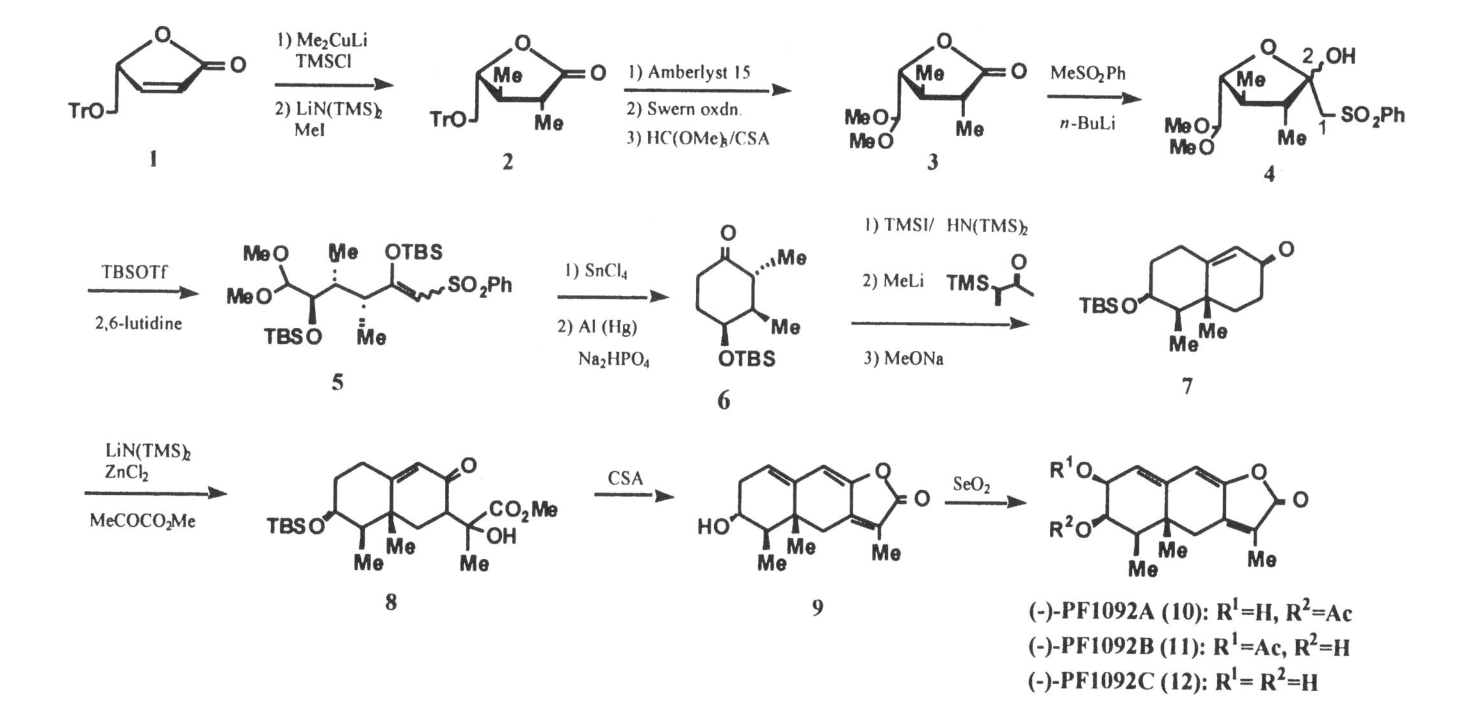
1
1) Me2CuLi TMSCl
2) LiN(TMS)2 MeI
2
1) Amberlyst 15
2) Swern oxdn.
3) HC(OMe)3/CSA
3
MeSO2Ph
n-BuLi
4
TBSOTf
2,6-lutidine
5
1) SnCl4
2) Al (Hg) Na2HPO4
6
1) TMSI/ HN(TMS)2
2) MeLi
3) MeONa
7
LiN(TMS)2 ZnCl2
MeCOCO2Me
8
CSA
9
SeO2
(-)-PF1092A (10): R1=H, R2=Ac
(-)-PF1092B (11): R1=Ac, R2=H
(-)-PF1092C (12): R1= R2=H

Scheme 1

with methyl pyruvate to give **8** quantitatively as a diastereomeric mixture. Closure to the desired lactone **9** was affected upon heating **8** with CSA. Finally, stereospecific SeO_2 oxidation of **9** with the aid of the hydroxy group afforded the *cis* diol **12**, identical with the natural product (-)-PF1092C (**12**) in all respects (*3*).

Since (-)-PF1092C (**12**) has already been transformed into (-)-PF1092A and B (**10** and **11**) by selective acetylation, the synthesis of **12** constitutes the completion of the total synthesis of **10** and **11** (*3*).

3. Total Synthesis of Glyoxalase I Inhibitor and Its Precursor KD16-U1

A glyoxalase I inhibitor (**19**) was isolated in 1975 from the culture broth of *Streptomyces griseosporeus* by Umezawa and co-workers. The absolute structure was determined by chemical studies and X-ray analysis (*5*). Its precursor, (-)-KD16-U1 (**18**), had been already isolated in 1974 from the culture broth of *Streptomyces filipinensis* by a chemical screening method developed in our laboratories (*6*), and converted to the aforementioned glyoxalase I inhibitor by treatment with crotonic acid and BF_3-Et_2O (*7*). The glyoxalase system, which consists of glyoxalase I, glyoxalase II and reduced glutathione, catalyzes the conversion of α-keto aldehydes to α-hydroxy acids. The glyoxalase I inhibitor (**19**) has also been reported to exhibit antitumor activities. The structures and bioactivities of these compounds **18** and **19** have attracted our attention because of our program in developing novel methodology for the preparation of densely functionalized carbocycles from carbohydrates. The first synthesis was achieved by Vasella et al. in which methyl α-D-glucopyranoside was effectively used as starting material (*8*). As mentioned in the synthesis of PF1092s (**10**– **12**) (*3*), the $SnCl_4$-promoted aldol-like cyclization of phenylsulfonyl enol silyl ethers containing a dimethyl acetal has been explored extensively in our laboratories. This transformation is ideally suited to the synthesis of carbocycle-containing natural products and carba-sugars, since the core skeleton arises after appropriate replacement of the phenylsulfonyl group. Accordingly, the novel synthesis of (-)-glyoxalase I inhibitor (**19**) and its precursor, (-)-KD16-U1 (**18**) has been accomplished by the similar manner (*9*) (Scheme 2).

The key step was the introduction of the hydroxymethyl group onto the α-phenylsulfonyl cyclohexenone **16** through the Michael type addition of tributylstannyl lithium followed by trapping with formaldehyde and desulfonylation.

The cyclohexenone **16** would arise from the enol silyl ether containing the dimethyl acetal **15**, which originates from one-step opening of the phenylsulfonylmethyl furanose **14**. Thus, the starting material simplifies to commercially available D-ribonic acid γ-lactone.

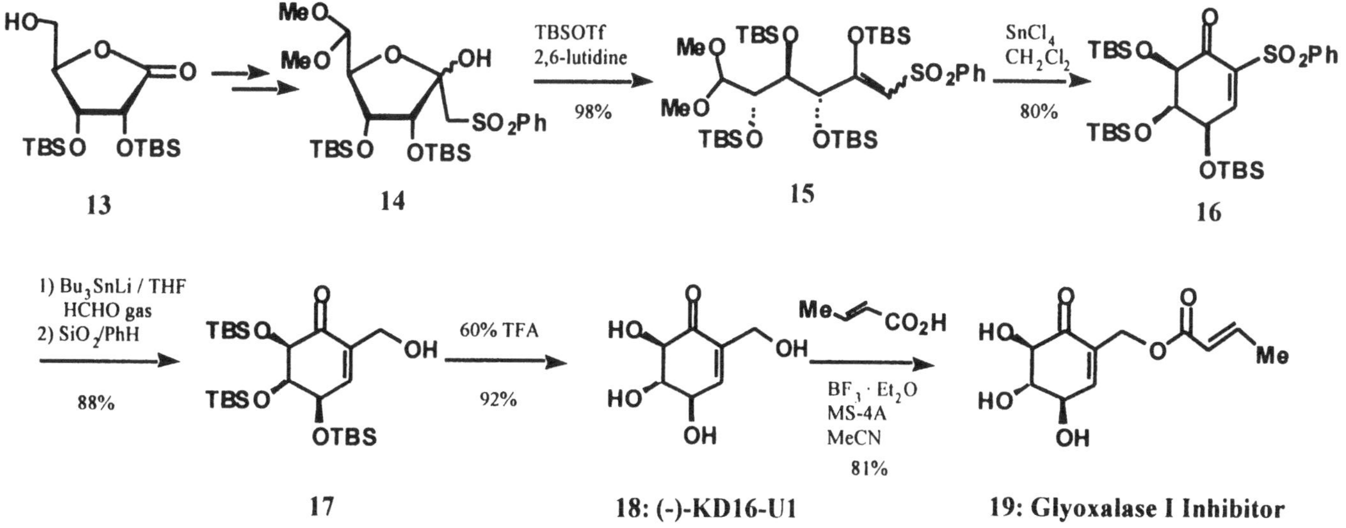

13 → 14

TBSOTf, 2,6-lutidine, 98%

15

$SnCl_4$, CH_2Cl_2, 80%

16

1) Bu_3SnLi / THF, HCHO gas; 2) SiO_2/PhH, 88%

17

60% TFA, 92%

18: (-)-KD16-U1

Me–CH=CH–CO_2H, $BF_3 \cdot Et_2O$, MS-4A, MeCN, 81%

19: Glyoxalase I Inhibitor

Scheme 2

In practice, the silylated lactone **13** was converted into **14** by acetal formation followed by reaction with lithiated methyl phenyl sulfone (*9*). Compound **14** was silylated to produce, as expected, the labile enol silyl ether **15** having a simultaneously silylated hydroxy group. The $SnCl_4$-promoted cyclization of **15** resulted in the formation of the cyclohexenone **16**. Trapping of the intermediary β-tributylstannyl sulfone with formaldehyde-followed addition of tributylstannyl-lithium to 16. This reaction gave an adduct which, upon treatment with silica gel, was converted through simultaneous elimination of the phenylsulfonyl and tributylstannyl groups to the desired α-hydroxymethyl-cyclohexenone **17**. De-*O*-silylation with 90% TFA afforded **18**, which was identical with the natural (-)-KD16-U1 (**18**) in all respects (*9*).

The synthetic (-)-KD16-U1 (**18**) was treated with crotonic acid and BF_3-Et_2O, as previously reported in our laboratories (*7*), to give the selectively acylated product **19** identical with the natural glyoxalase I inhibitor (**19**).

4. Novel Synthesis of Natural Pseudo-aminosugars, (+)-Valienamine and (+)-Validamine

(+)-Valienamine (**25**) and (+)-validamine (**29**) have been found to be key components for biological activities in pseudo-aminosugars and pseudo-oligosaccharides such as validamycins, acarbose, and trestatines (*10*). Both pseudo-aminosugars **25** and **29** were also isolated from the fermentation broth of *Streptomyces hygroscopicus* subsp. *limoneus* IFO 12703 to show some biological activities.

Few syntheses of optically active compounds **25** and **29** have been reported by using L-quebrachitol, (-)-quinic acid, and D-glucose derivatives, although the racemates of **25** and **29** have been synthesized in a variety of methodologies.

As mentioned in the synthesis of the glyoxalase I inhibitor (**19**) and its biosynthetc precursor **18**, we have extensively developed the one-step opening of a furanose ring containing a phenylsulfonylmethyl group followed by aldol condensation as a general method for the construction of optically active carbasugars (*9*).

Now, both the utility and the versatility of our method are demonstrated in the stereoselective synthesis of natural (+)-valienamine (**25**) and (+)-validamine (**29**) (*11*) (Scheme 3). Furthermore, the anchor effect of an amino group will be described in the stereoselective hydrogenation of the olefin of**25** to give **29** (Fig. 1).

The starting compound **20**, which was prepared from D-xylose by bromine oxidation and tritylation, was converted into **21** by the similar procedures with the synthesis of **17**.

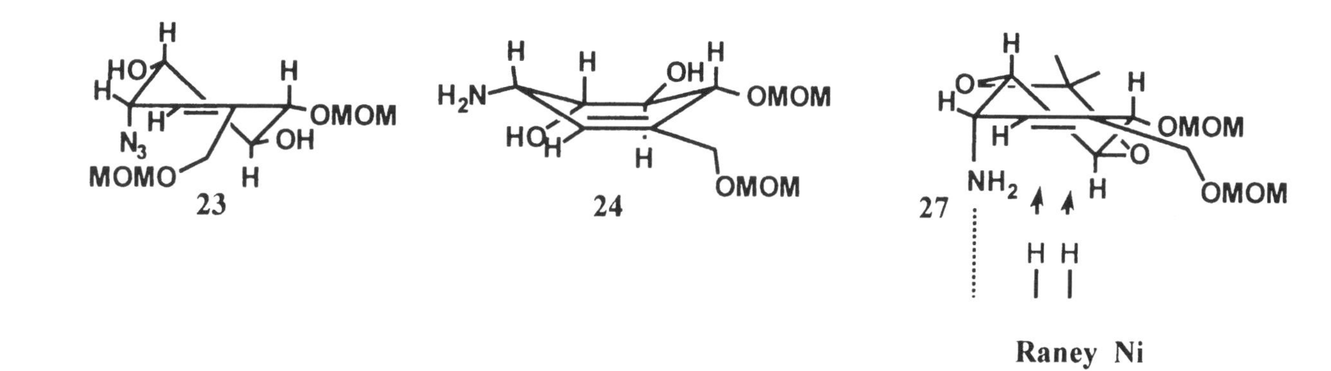

Fig. 1. Conformations of key intermediates and the anchor effect of the amino group in **27** over Raney Ni.

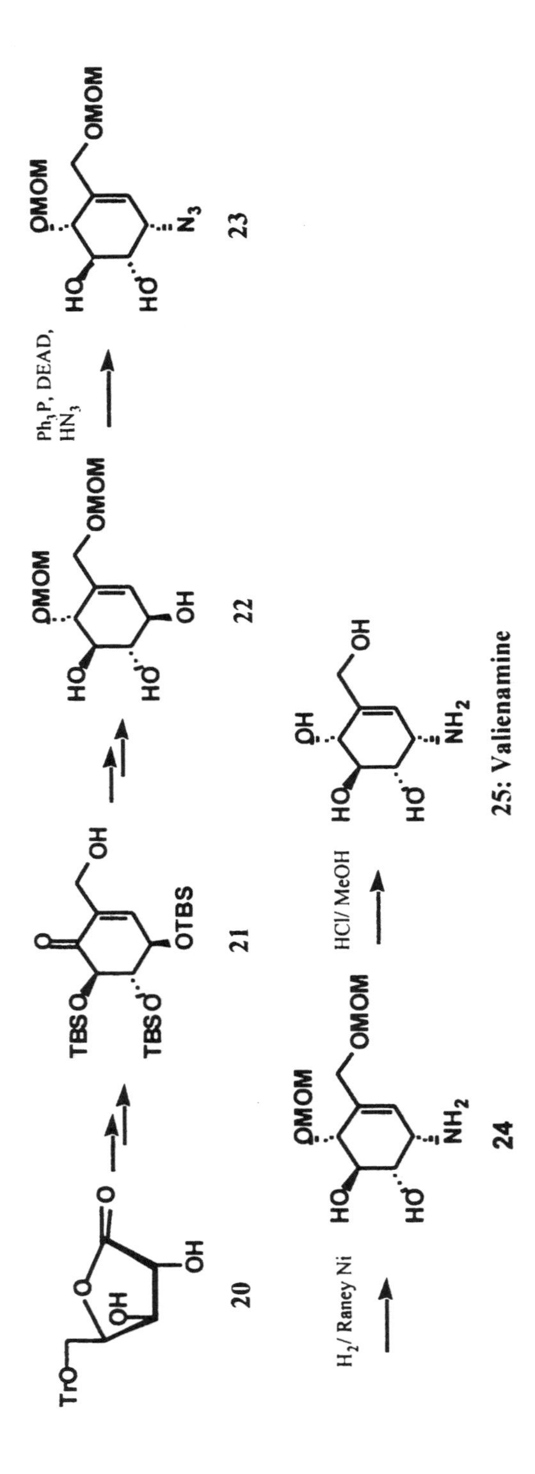

TrO
OH
O
OH
20
TBSO
TBSO
O
OTBS
OH
21
HO
HO
OMOM
OMOM
OH
22
Ph3P, DEAD,
HN3
HO
HO
OMOM
OMOM
N3
23
H2/ Raney Ni
HO
HO
OMOM
OMOM
NH2
24
HCl/ MeOH
HO
HO
OH
OH
NH2
25: Valienamine

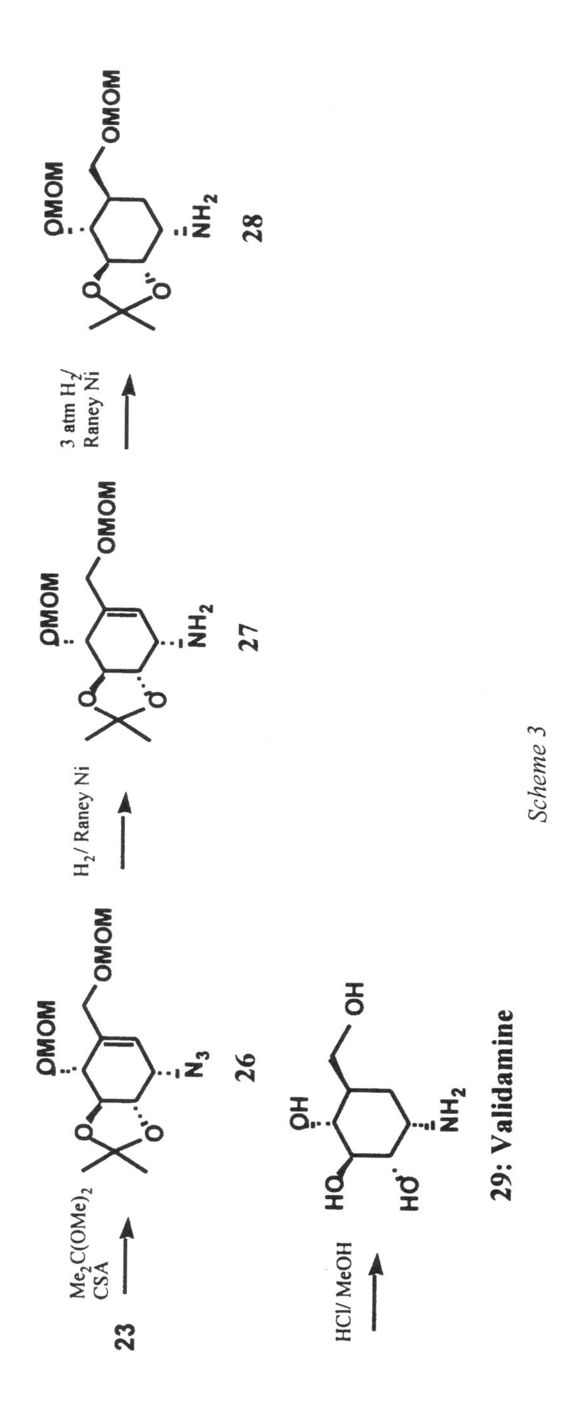

23
$Me_2C(OMe)_2$
CSA
OMOM
OMOM
N_3
26
H_2/ Raney Ni
OMOM
OMOM
NH_2
27
3 atm H_2/
Raney Ni
OMOM
OMOM
NH_2
28
HCl/ MeOH
HO
HO
OH
OH
NH_2
29: Validamine

Scheme 3

Stereoselective reduction of the carbonyl group in **21** by $Zn(BH_4)_2$ in ether to give the α-alcohol was followed by exchange of the protecting groups to give the properly protected alcohol **22**.

Although **22** possessed three hydroxy groups, the allyl hydroxy group at C-1 was expected to be more reactive than others.

As expected, Mitsunobu inversion of the allyl alcohol **22** using HN_3 gave predominantly the α-azide **23** (*12*). Mild hydrogenation of **23** with 1 atm of hydrogen over Raney Ni produced the corresponding amino compound **24** in a quantitative yield without any significant reduction of the olefin.

Deprotection of **24** with methanolic hydrogen chloride gave the hydrochloride of (+)-valienemine (**25**), which was chromatographed on Dowex 1X2 (OH-type) with water to provide, after recrystallization from a little of water, needles of the free base **25** as a monohydrate. Both the hydrochloride and the free base of **25** were identical in all respects with the authentic samples of the natural product (*11*).

On the final stage to synthesize validamine (**29**), extensive efforts were directed toward stereoselective hydrogenation of the olefin **24** that would ensure the configuration of the hydroxymethyl group at C-5.

Unfortunately, catalytic hydrogenation of **24** either on Raney Ni or Pd-C gave an approximately 1: 1 mixture of the diastereomers due to the C-5.

The ^{1}H-NMR studies of **23** and **24** indicated that their conformations were different (Fig. 1). Compound **23** adopts the usual half-chair form with the *quasi*-axial azido group, while the amino compound **24** exists in the boat-like form with the *quasi*-equatorial amino group. The latter form **24** might be due to the interaction such as hydrogen bonding between the C-1 amino and C-2 hydroxy groups.

The *quasi*-axial amino group or hydroxy group of methylcyclohex-2-enylamines or 2-enols has been known to act as the anchor toward the surface of Raney Ni on catalytic hydrogenation to give preferentially the *trans* isomer (*13*).

Accordingly, the *quasi*-equatorial amino group of **24** could not participate in the anchor effect for stereoselective hydrogenation.

We expected that the conformation of the acetonated derivatives **26** and **27** would be much more rigid than **24** to keep the half-chair form having the *quasi*-axial amino group. Mild catalytic hydrogenation of **26** gave quantitatively the corresponding amino compound **27**.

As a mixture of dioxane and H_2O was used for catalytic hydrogenation, the ^{1}H-NMR spectra of **24** and **27** were measured in dioxane-d_6 and D_2O to support that existed in the half-chair form with the *quasi*-axial amino group (Fig. 1). The *quasi*-axial amino group of **27** was expected to assist the anchor effect giving the desired 1,5-*trans* isomer **28**.

As expected, catalytic hydrogenation of **27** over Raney Ni in a mixture of dioxane and H_2O gave, after evaporation of the solvent, the *trans* isomer **28** as a single product in a quantitative yield. Direct hydrogenation of **26** to **28** was also achieved in a quantitative yield with 3 atm of hydrogen on Raney Ni.

Acidic deprotection of **28** gave quantitatively the hydrochloride of (+)-validamine (**29**), which was chromatographed on Dowex 1X2 (OH-type) with water to yield the free base of **29**. Both the hydrochloride and the free base of **29** were identical in all respects with the authentic samples of the natural product (*11*).

In summary, the novel synthesis of (+)-valienamine and (+)-validamine has been accomplished by our synthetic strategies for the constructing of carbasugars. The absolutely stereoselective hydrogenation of **26** and **27** to give (+)-validamine (**29**) is particularly noteworthy.

5. The First Total Synthesis of Pyralomicin 1c and 2c

Pyralomicins 1c and 2c (**36** and **38**) have been isolated from culture broth of *Microtetraspora spiralis* as novel antibiotics including antitumor activities. Structurally, **36** and **38** are endowed with the 5-hydroxy-8-methyl- [1]-benzopyrano[2,3-b]pyrrol-4-(1*H*)-one structure **32** as a common core binding a carba sugar and a sugar moiety, respectively.

The first total synthesis of pyralomicin 1c and 2c (**36** and **38**) has been effectively accomplished in our laboratoriess (*15*, *16*) (Scheme 4).

Since pyralomicinone (**32**) has been synthesized from pyrrole and 2,4-dihydroxytoluene derivatives (for examples: **30** and **31**), the first aim in the synthesis of pyralomicin 1c (**36**) is the effective construction of the carba sugar moiety **35** (*15*).

We expected the regio- and stereoselective connection of **35** with pyralomicinone (**32**) to be controlled under Mitsunobu conditions with inversion.

Furthermore, it was anticipated that the carba sugar **35** would be synthesized by the similar strategies as developed by us in the syntheses of glyoxalase I inhibitor (**19**) and its precursor, KD16-U1 (**18**) (*9*).

The starting material in this synthesis was L-arabinonic acid γ-lactone **33**, which was readily derived from L-arabinose by tritylation and bromine oxidation (*11*, *15*). Conversion of **33** to **34** was carried out by the similar procedures with the preparation of **17** and **21**.

Stereoselective reduction of the carbonyl group of **34** was examined in a variety of conditions, and the best result was realized by using $NaBH_4$ and $CeCl_3 \cdot 7H_2O$ to give the desired α-alcohol in 69%. This was protected with methoxymethyl group followed by de-*O*-silylation to give quantitativelythe triol **35**. Although **35** possessed three free hydroxy groups, the allyl hydroxy group at C-1 was expected to be more reactive than others.

With pyralomicinone (**32**) and the alcohol **35** in hand, we turned to their connection. Both components **32** and **35** were coupled under modified

30

31

32

33

34

35

32
$Bu_3P=CHCN$

36: Pyralomicin 1c

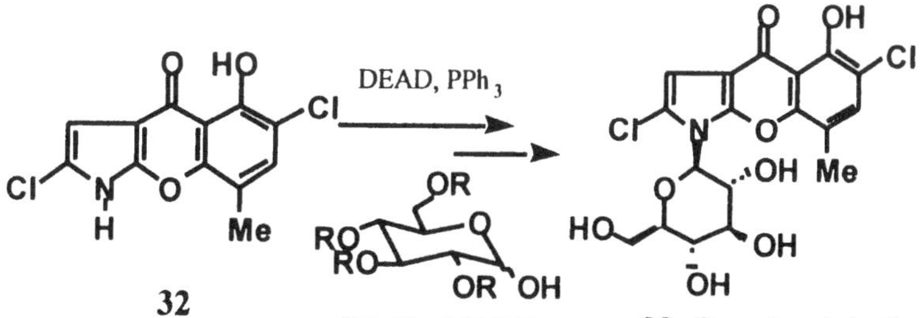

32 | 37: R=MOM | 38: Pyralomicin 2c

Scheme 4

Mitsunobu's conditions using a novel reagent, *n*-Bu_3P=CHCN to give predominantly the desired N-C product with inversion (*17*), which was deprotected under acidic conditions to give pyralomicin 1c (**36**). This was identical with the natural product in all respects, completing the first total synthesis. As expected, the by-products, which would result from the reaction of other hydroxy groups with 32, were not significantly observed.

Next, pyralomicin 2c (**38**) was synthesized from **32** and **37** (*16*). The glucosyl donor **37** was prepared from benzyl α-D-glucopyranoside by methoxymethylation followed by hydrogenolysis. The stereoselective *N*-glycosylation of **37** with **32** was effectively accomplished by using Mitsunobu conditions to give **38**, after acidic deprotection. This was identical in all respects with the natural product **38** (*16*).

The total synthesis of pyralomicin 1c (**36**) and 2c (**38**) indicated that Mitsunobu conditions were useful for the stereoselective construction of N-C bonds.

6. The First Total Synthesis and Mode of Action of N-Methyl-D-aspartate Receptor Antagonists, ES-242s

A bioxantracene (-)-ES-242-4 (**46**) was isolated from the culture broth of *Verticillium* sp. in 1992 as one of eight antagonists for the *N*-methyl-D-aspartate (NMDA) receptor (*18*). These novel natural products are reported to inhibit the [^{3}H]thienyl cyclohexylpiperidine binding to rat crude synaptic membranes, and therefore, are of potential therapeutic interest for the treatment of neurodegenerative diseases. (-)-ES-242-4 (**46**) is structurally remarkable having an axially chiral binaphthalene core that is adorned with two pyrans of the same absolute chirality. Our interest in the construction of densely functionalized naphthopyran ring systems, *via* tandem Michael-Dieckmann reactions, promoted us to attempt the first stereocontrolled total synthesis of (-)-ES-242-4 (**46**) (*19*).

To this end, the naphthopyran derivative **44** was our first target, which could be derived from the α,β-unsaturated lactone **40** and the *o*-methylbenzoate **41** through Michael and Dieckmann reactions. It was expected that the pivotal conversion of a monomer **44** to a dimer **46** could be accomplished by oxidative coupling (Scheme 5).

On one hand, the α,β-unsaturated lactone **40**, which was derived from di-*O*-acetyl-L-rhamnal (**39**) according to reported procedures, was submitted to Mitsunobu inversion with HCO_2H, followed by hydrolysis and methoxymethylation to afford **40**. On the other hand, the *o*-methylbenzoate **41** was obtained from 3,5-dihydroxytoluene under the protocols described by Solladié.

Addition of lithiated **41** to **40** was followed by Dieckmann reaction to provide a single product **42** as expected from *trans* addition to the C-4 *O*-MOM group. Aromatization of **42** was followed by *O*-benzylation to give **43**. Hydride reduction of **43** to the lactol was followed by treatment with Et_3SiH and TFA. The pyran **44** was obtained. Oxidative dimerization of **44** was examined under several conditions with a variety of metals such as Fe(II), Mn(II) and Cu(II). The best result was realized by the protocols reported by Noji, Nakajima and Koga using CuCl(OH)·TMEDA (*20*), which was prepared from CuCl and TMEDA under oxygen. The diastereomeric mixture of **45** was produced as a stable intermediate (IR [KBr] 1648 cm^{-1}). Finally, **45** was aromatized with aq. NaOH followed by acid hydrolysis to remove the *O*-MOM group. Expectedly, two atropisomers were produced and isolated by silica gel column chromatography to give **46** and **47** in 37% and 38% overall yields, respectively. The former **46** was identical in all respects with an authentic sample of the natural (-)-ES-242-4 (*19*).

Furthermore, the diastereomeric analogs (**49** and **50**) of ES-242-4 (**46**) were synthesized from **39** by the similar synthetic strategies but without isomerization of the C4 hydroxy group to understand the structure-activity relationships. Similarly, Michael-Dieckmann type reaction of **48** with **41** gave the tricyclic compound, which was converted into the atropisomers **49** and **50** (*21, 22*).

The absolute and atropisomeric structures of all isomers **46, 47, 49** and **50** were determined by their chemical derivation and the X-ray crystallographic analysis of *O*-benzyl derivative of **49** (*23*).

Finally, the structure-activity relationships were disclosed as follows (*23*): Two hydroxy groups at C-4 and C-4' in **46** (ES-242-4) and **49** are observed to be far apart, while two hydroxy groups in **47** and **50** are close together. The shorter distance between these two hydroxy groups may be responsible for the stronger inhibitory activities against [^{3}H]MK-801 binding to the NMDA receptor (*23*). Namely, **47** and **50** showed stronger activities than **46** and **49**, suggesting that the appearance of their activities may be attributed to the intramolecular metal chelation formation between their two hydroxy groups.

7. The First Total Synthesis of Natural (-)-Tetracycline

For almost half a century, tetracycline (**68**) has been well-known as a major antibiotic from the viewpoint of its unique structural features as well as antibacterial activities (*24*). The total synthesis of tetracycline families was initiated by Woodward's 6-demethyl-6-deoxytetracycline synthesis in 1962 (*25*), followed by Muxfeldt's terramycin synthesis in 1968 (*26*), and culminated by Stork's 12a-deoxytetracycline synthesis in 1996 (*27*). However, all these syntheses have been accomplished only in racemic forms. The total synthesis of

AcO, Me, O, OAc

39

MOMO, Me, O, O

40

OMe, CO_2Me, MeO, Me

41

LDA /THF

MeO, OH, O, O, MeO, H, Me, OMOM

42

MeO, OBn, O, O, MeO, Me, OMOM

43

MeO, OH, O, MeO, Me, OMOM

44

CuCl(OH)•TMEDA

O_2 /CH_2Cl_2

92%

MeO, O, O, MeO, Me, OMOM, OMOM, Me, MeO, O, MeO, O

45

1) 1M-NaOH MeOH

2) AcCl /MeOH/ dioxane

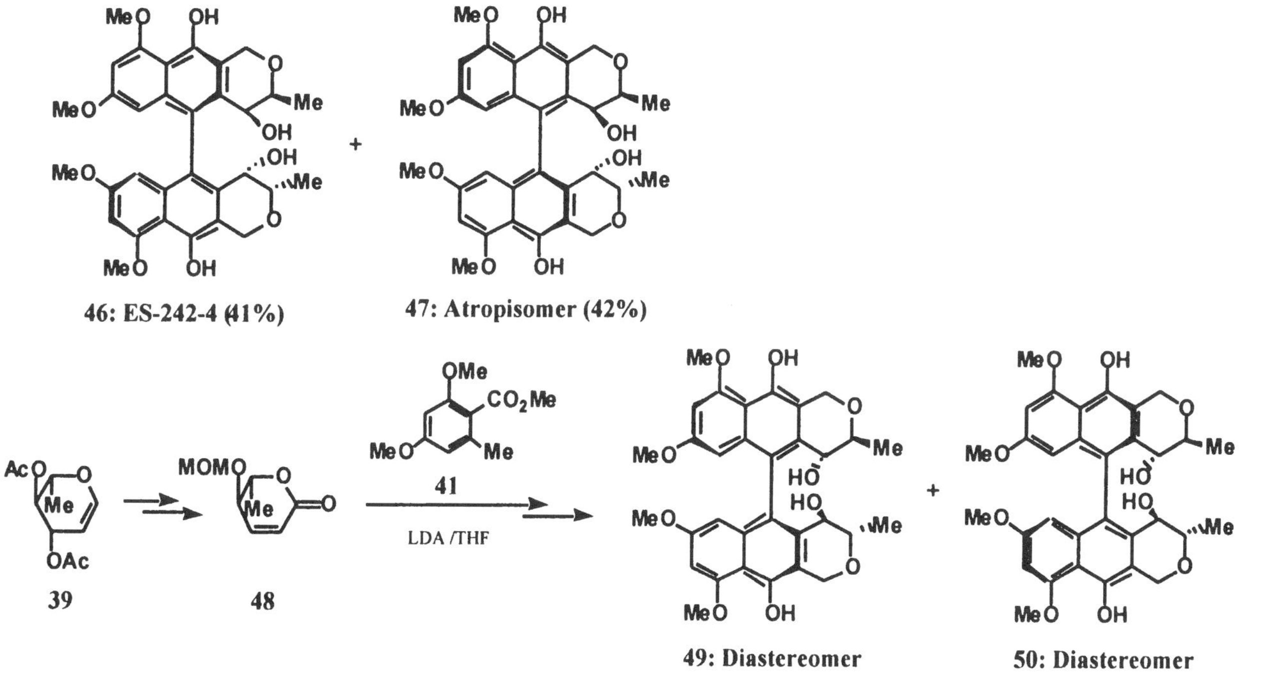

MeO OH MeO OH
MeO Me OH OH Me MeO
MeO OH

46: ES-242-4 (41%)

+

47: Atropisomer (42%)

AcO O Me OAc

39

MOMO O Me O

48

OMe CO_2Me MeO Me

41

LDA /THF

49: Diastereomer

+

50: Diastereomer

Inhibitory Activities [IC_{50} (μM/ml)] against the Binding of [^{3}H]MK-801

Compounds	46	47	49	50
Activities	40	14	ÂÑ200	0.4

Scheme 5

natural (-)-tetracycline (**68**) remained an unanswered challenge, despite the remarkable achievements as described above.

Very recently, the first total synthesis of (-)-tetracycline (**68**) has been completed in our laboratories (*28*) by using D-glucosamine as a chiral starting material, which stereospecifically constructs the densely and sensitively functionalized A ring (Scheme 6).

From the retrosynthetic perspective (Fig. 2), the tetracyclic structure is expected to be accessible by tandem Michael-Dieckmann type reaction of **59** with **60**. The suitably substituted chiral intermediate **59** would be synthesized by Diels-Alder reaction of the cyclohexenone **57** and the silyloxybutadiene **58**. The regio- and stereoselectivities are established as a consequence of the dienophile geometry according to Gleiter's theory (*29*). Compound **57** could be obtained from **51** through Ferrier reaction of **54**.

As a viable synthetic relay from anhydrotetracycline (**66**) to tetracycline (**68**) has been reported by Wasserman and Scott via a two-step hydration at the 5a, 6-position (*30*), **66** was our first target. A reliable 12a-hydroxylation is required for the synthesis of **66**, although evidence of such hydroxylation has been reported (*27, 31*).

The starting **51**, which was prepared from D-glucosamine, was converted into the olefin **52** by selective silylation, oxidation and Wittig olefination (Scheme 2). After de-*O*-silylation of **52**, the resulting alcohol was led to the selenide **53**. Treatment of **53** with borane followed by H_2O_2 oxidation gave stereoselectively the alcohol by simultaneous formation of a new olefin group, which was benzylated to **30**. This was submitted to Ferrier reaction (*32*) with $HgCl_2$ to give the cyclohexanone **55**. The [4+2] cycloaddition of **56**, which was derived from **55** by dehydration, with the butadiene **58** did not proceed because of the steric repulsion. Therefore, **55** was epimerized at C2 and dehydrated to the isomer **57**. The α-hydroxymethyl group was an important factor for the stereospecific introduction of the hydroxy group at 12a in **63** and **64**. This cycloaddition with **58** in the presence of 2,6-di-*tert*-butyl-4-methylphenol (DBMP) proceeded from the β-face of **57** regio- and stereoselectively as expected. This highly stereoselective reaction gave a labile adduct, which upon acidic oxidation was transformed to the α,β-unsaturated ketone **59**. The tandem Michael-Dieckmann type reaction of **59** with the isobenzofuranone **60** gave the tetracyclic compound, which was in turn aromatised to **61** in high yield.

After selective de-*O*-benzylation of **61** with BBr_3 (Scheme 3), the alcohol was converted into **62** by exchange of the *N*-protecting group followed by *O*-methylation of the enol Treatment of **62** with Br_2 gave stereoselectively the bromide **63**. The opening of the pyran ring was examined under a variety of conditions. PCC-PDC oxidation of the alcohol of **63** was found to give the C12a alcohol **64** followed by β-elimination and oxidative opening of the pyran. Compound **64** was transformed to the nitrile **65** by our newly developed method.

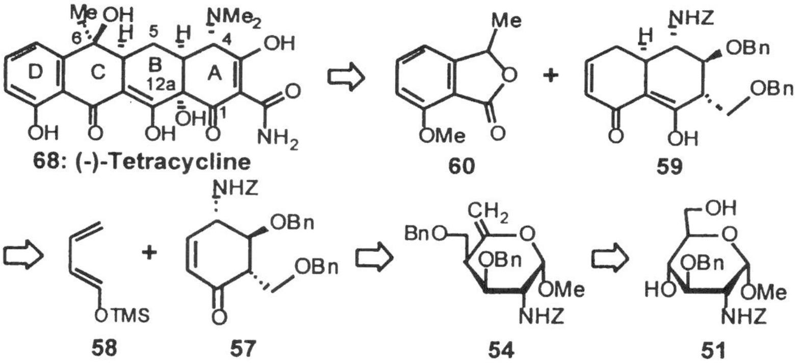

Fig. 2 Retrosynthetic Approach to Tetracycline

51 → 52 → 53

1) $BH_3 \cdot THF$, H_2O_2/NaOH
2) BnBr

54

$HgCl_2$

55

55 → 56

57

58, 170°C

CrO_3, H_2SO_4

59

60, LDA

$SOCl_2$, TEA

61 → 62

Br_2, $(Bu_3Sn)_2O$

63

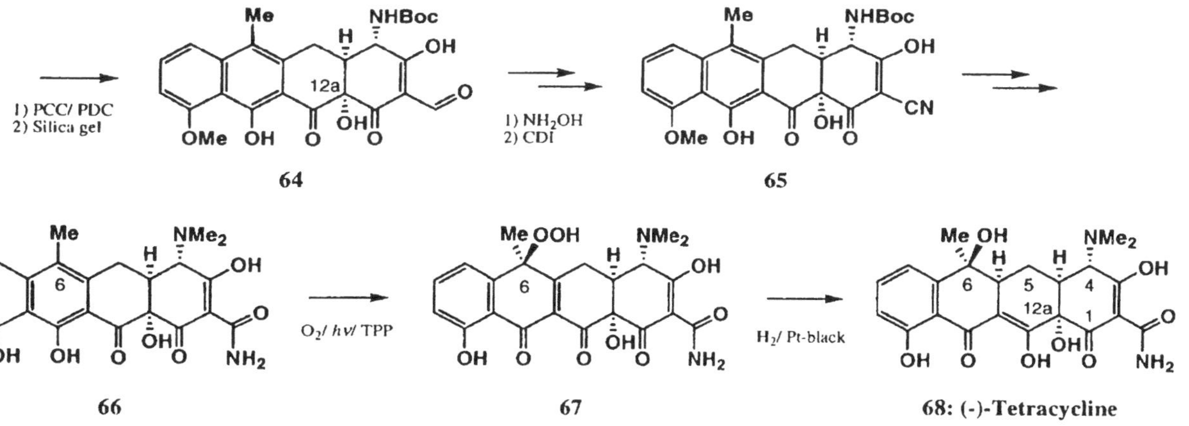
1) PCC/ PDC
2) Silica gel
Me
NHBoc
H
OH
12a
OMe OH O OH O
64
1) NH$_2$OH
2) CDI
Me
NHBoc
H
OH
CN
OMe OH O OH O
65
Me
NMe$_2$
H
OH
6
OH OH O OH O NH$_2$
66
O$_2$/ $h\nu$/ TPP
Me OOH
NMe$_2$
H
OH
6
OH O O OH O NH$_2$
67
H$_2$/ Pt-black
Me OH
NMe$_2$
H H
OH
6 5 4
12a 1
OH O OH OH O NH$_2$
68: (-)-Tetracycline

Scheme 6

Hydrolysis of **65** to give the amide with concomitant removal of the *N*-Boc group was followed by *N*-dimethylation and de-*O*-methylation to produce anhydrotetracycline (**66**). This was identical with a naturally derived sample in all respects.

The final stage was to introduce stereoselectively the hydroxy group into the C6 position according to the reported procedures (*30*). By photooxidation of **66**, the corresponding C-6 peroxide **67** was obtained. The successive hydrogenolysis on Pd-C gave no significant product (*27, 30*), while the desired reduction proceeded smoothly on Pt black to give (-)-tetracycline (**68**) in a fairly good yield, which was neutralized with HCl in MeOH to give the hydrochloride. This was identical with the hydrochloride of natural (-)-tetracycline in all respects, completing the first total synthesis (*28*).

8. Conclusion

The total synthesis of medicinally useful natural products was accomplished by using carbohydrates to determine their absolute structures and to illustrate the usefulness of carbohydrates as chiral sources, and the analogs of the natural products were developed to further clarify the structure - activity relations.

Acknowledgments: The author is grateful to Advanced Research Institute for Science and Engineering, Waseda University, and High-Tech Research Center Project the Ministry of Education, Science, Sports and Culture for the generous support of our program. Financial support by the Grant-in-Aid for Specially Promoted Research from the Ministry of Education, Science, Sports and Culture is also gratefully acknowledged.

References

1. Tatsuta, K. *J. Synth. Org. Chem. Jpn.*, **1997**, *55*, 970.
2. Tabata, Y.; Hatsu, M.; Kurata, Y.; Miyajima, K.; Tani, M.; Sasaki, T.; Kodama, Y.; Tsuruoka T.; Omoto, S. *J. Antibiot.*
3. Tatsuta, K.; Yasuda, S.; Kurihara, K.; Tanabe, K.; Shinei, R.; Okonogi, T. *Tetrahedron Lett.* **1997**, *38*, 1439.
4. Stork, G.; Singh, J. *J. Am. Chem. Soc.*, **1974**, *96*, 6181.
5. Chimura, H.; Nakamura, H.; Takita, T.; Takeuchi, T.; Umezawa, H.; Kato, K.; Saito, S.; Tomisawa, H.; Iitaka, Y. *J. Antibiot.* **1975**, *28*, 743.
6. Tatsuta, K.; Tsuchiya, T.; Mikami, N.; Umezawa, S.; Umezawa, H.; Naganawa, H. *J. Antibiot.* **1974**, *27*, 579.
7. Umezawa, S.; Tatsuta, K.; Takeuchi, T.; Umezawa, H.; Takita, T. *Chem. Abstr.* **87**, 184080b (1977).

8. Mirza, S.; Molleyres, L.-P.; Vasella, A. *Helv. Chim. Acta*, **1985**, *68*, 988.
9. Tatsuta, K.; Yasuda, S.; Araki, N.; Takahashi, M. *Tetrahedron Lett.* **1998**, *39*, 401.
10. Fukase, H. *Yuki Gosei Kagaku Kyokaishi*, **1997**, *55*, 920.
11. Tatsuta, K.; Mukai, H.; Takahashi, M. *J. Antibiot.* **2000**, *53*, 430.
12. Tatsuta, K.; Yasuda, S.; Araki, N.; Takahashi, M.; Kamiya Y. *Tetrahedron Lett.* **1998**, *39*, 401.
13. Sugi, Y.; Mitsui, S. *Chem. Lett.* **1974**, *1974*, 577.
14. Kawamura, N.; Nakamura, H.; Sawa, R.; Takahashi, Y.; Sawa, T.; Naganawa, H.; Takeuchi, T. *J. Antibiot.* **1997**, *50*, 147.
15. Tatsuta, K.; Takahashi, M.; Tanaka, N. *J. Antibiot.* **2000**, *53*, **88**.
16. Tatsuta, K.; Takahashi, M.; Tanaka, N. *Tetrahedron Lett.* **1999**, *40*, 1929.
17. Tsunoda, T.; Ozaki, F.; Shirakata, N.; Tamaoka, Y.; Yamamoto, H.; Ito, S. *Tetrahedron Lett.* **1996**, *37*, 2463.
18. Toki, S.; Ando, K.; Kawamoto, I.; Sano, H.; Yoshida, M.; Matsuda, Y. *J. Antibiot.* **1992**, *45,* 1047.
19. Tatsuta, K.; Yamazaki, T.; Mase, T.; Yoshimoto, T. *Tetrahedron Lett.* **1998**, *39*, 1771.
20. Noji, M.; Nakajima, M.; Koga, K. *Tetrahedron Lett.* **1994**, *35*, 7983.
21. Tatsuta, K.; Yamazaki, T.; Yoshimoto, T. *J. Antibiot.* **1998**, *51*, 383.
22. Tatsuta, K.; Nagai, T.; Mase, T.; Yamazaki, T.; Tamura, T. *J. Antibiot.* **1999**, *52*, 422.
23. Tatsuta, K.; Nagai, T.; Mase, T.; Tamura, T.; Nakamura, H. *J. Antibiot.* **1999**, *52*, 433.
24. Tatsuta, K.; Miura, S.; Gunji, H. *Bull. Chem. Soc. Jpn.*, **1997**, *70*, 427.
25. Conover, L. H.; Butler, K.; Johnston, J. D.; Korst, J. J.; Woodward, R. B. *J. Am. Chem. Soc.*, **1962**, *84*, 3222.
26. Muxfeldt, H.; Hardtmann, G.; Kathawala, F.; Vedejs, E.; Mooberry, J. B. *J. Am. Chem. Soc.*, **1968**, *90*, 6534.
27. Stork, G.; La Clair, J. J.; Spargo, P.; Nargund, R. P.; Totah, N. *J. Am. Chem. Soc.*, **1996**, *118*, 5304.
28. Tatsuta, K.; Yoshimoto, T.; Gunji, H.; Okado, Y.; Takahashi, M. *Chemistry Lett.* **2000**,
29. Gleiter, R.; Böhm, M. C. *Pure Appl. Chem.*, **1983**, *55*, 237.
30. Wasserman, H. H.; Lu, T. -J.; Scott, A. I. *J. Am. Chem. Soc.*, **1986**, *108*, 4237.
31. Gurevich, A. I.; Karapetyan, M. G.; Kolosov, M. N.; Korobko, V. G.; Onoprienko, V. V.; Popravko, S. A.; Shemyakin, M. M. *Tetrahedron Lett.* **1967**, *1967*, 131.
32. Ferrier, R. J. *J. Chem. Soc., Perkin Trans. 1*, **1979**, 1979, 1455.

Chapter 10

Synthesis of Natural and Unnatural Products from Sugar Synthons

Minoru Isobe and Yoshiyasu Ichikawa

Laboratory of Organic Chemistry, School of Bioagricultural Sciences, Nagoya University, Chikusa, Nagoya 464–8601, Japan

We have developed two methods for the synthesis of natural and unnatural products from D-glucose. Enantio-and diastereo-switching method established a new strategy for the synthesis of four possible stereomers for natural products synthesis, and this powerful method was successfully applied to the synthesis of unnatural protein phosphatase inhibitors. The second synthetic method involved the preparation of the urea-glycosidic linkages for the synthesis of glycopeptide mimics.

Enantio-switching Method for the Synthesis of Natural and Unnatural Products from Sugar Synthons

"Carbohydrate synthons in natural product synthesis" which is the title of this special symposium has an inevitable problem to be solved, because we often need both enantiomers of synthons for the synthesis of optically active natural products, however, only one enantiomer of sugar synthon is available in many cases. We realized this problem during the retro-synthetic analysis of tautomycin **1** (TTM) as shown in Figure 1.

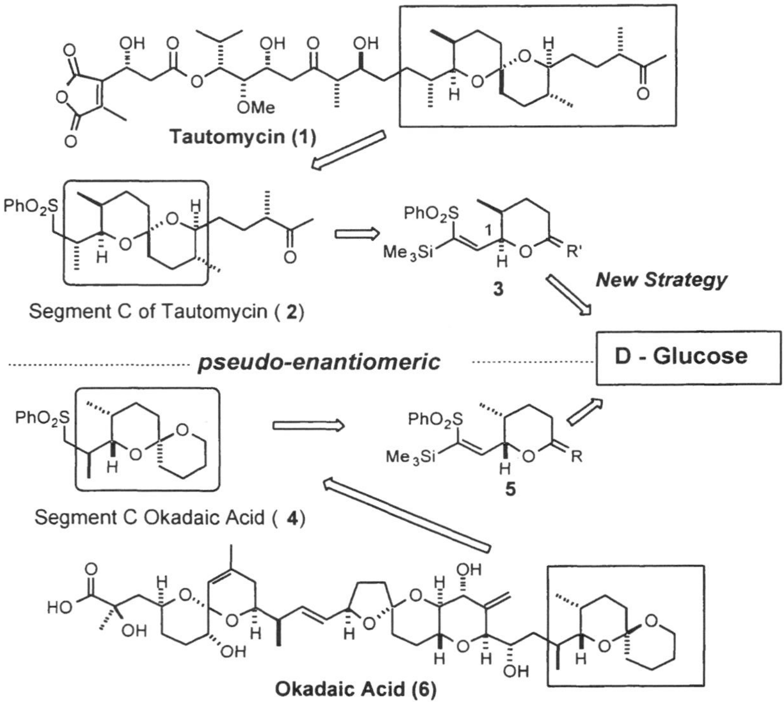

Figure 1. Retrosynthetic analysis of tautomycin

Since segment C of TTM **2** is structurally similar to Segment C of okadaic acid (OKA) **4**which had been prepared from D-glucose (*1*), we proposed a synthetic plan of Segment C of TTM **2** employing heteroolefin **3** which may also be derived from pyranose sugar derivative. But Segment C of TTM**2** is *pseudo-enantiomeric* form to its counterpart of Segment C of OA **4**; we recognized a problem associated with difficult availability of L-glucose. This problem led us to develop a new synthetic method "the *enantio-switching method*" which was applicable to both enantiomers starting from readily available D-glucose derivatives as shown in Figure 2 (*2*).

Enantio-switching Method for the Synthesis of Segment C of TTM

Enantio-switching method comprises *C*-glycosidation of alkyne, epimerization of its cobalt complex (*3*) and 1,2-asymmetricinduction *via* hetero conjugate addition to hetero olefins (*4*). Namely, *C*-glycosidation of silyl acetylene to the D-glucose derivatives **9** proceeds in a completely alpha-axial manner, and further transformation of the acetylenic group with bis-cobaltoctacarbonyl gave the bis-cobalthexacarbonyl complex **10**.

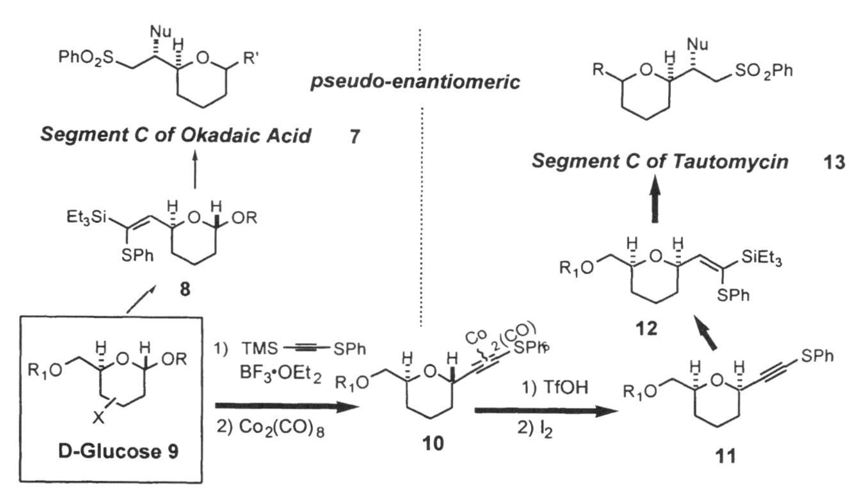

Figure 2. Enantio-switching method for the synthesis of Segment C of tautomycin

Epimerization of alpha-configuration of **10** into a β-equatorial configuration through acidic treatment and decomplexation by iodine provided the pseudo enantiomeric form of D-glucose derivative **11** (*5*). Finally, **11** is converted into the hetero olefin **12** which then received a variety of nucleophiles by chelation control to accomplish the synthesis of segment C of TTM **13**.

Enantio-switching and Diastereo-switching Method for Natural Products Synthesis

We have further developed this enantio-switching method for the synthesis of natural products as outlined in Figure 3. In this method, enantio-switching method was combined with diastereo-switching method which is based on α- and β-chelation control of the face selectivity of heteroolefin (*6*). Thus, *C*-glycosidation of D-glucose derivatives at C-1 position of pyranose and hydrosilylation provided the axial-hetero olefin **A,** which received α-chelation controlled heteroconjugate addition by alkyllithium to afford 1,2-*syn* isomer **C.** β-Chelation controlled addition of alkylmagnesium bromide to heteroolefin **A** provided 1,2-*anti* isomer **D** (*diastero-switching method*) (*7*). Enantio-switching method depicted in Figure 2 gave equatorial heteroolefin **B** which was transformed into pseudo-enantiomeric 1,2-*syn* and 1,2-*anti* isomers **E** and **F** by diastero-switching method. This strategy opened a completely stereo controlled fashion for the synthesis of four possible 1,2-*syn*- and *anti*-isomers starting from only one D-glucose derivative. Most notable point in this method is the newly generation of asymmetric centers in the acyclic portion using the pyranose as a chiral auxiliary. This method expanded the usefulness of sugar synthons for natural products synthesis and, in fact, applied to our study of the protein phosphatase inhibitor.

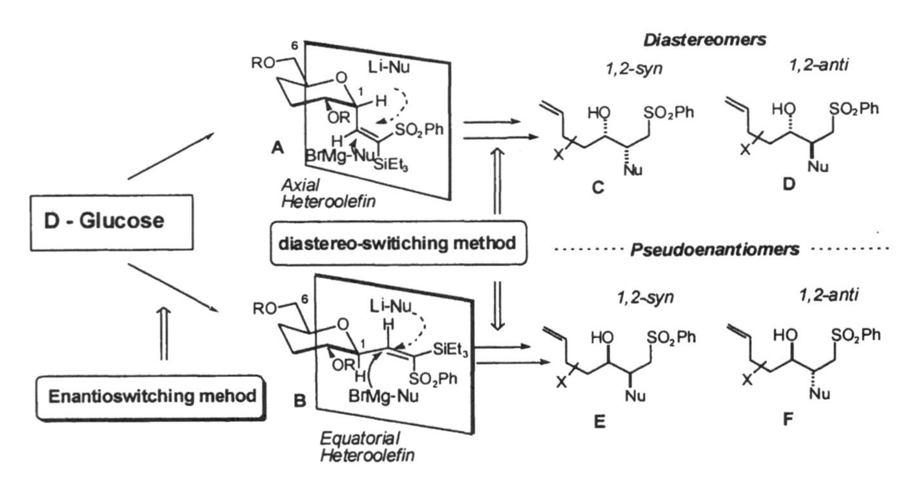

Figure 3. Enantioswitching and Diastereoswitching Method

Synthesis and Structural Recognition Studies of Protein Phosphatases Inhibitors

Our synthesis of OKA and TTM provided various opportunities for collaborating with biochemists, and this collaboration has given birth to our expanding interests toward structural recognition between protein phosphatases (PP) and inhibitors such as OKA or TTM (*8*). OKA has strong inhibitory effect to protein phosphatase Type 2A (PP2A), while TTM inhibits more effectively to protein phosphatase Type 1 (PP1) than type PP2A. We expected that structural difference of segment C portion of OKA and TTM determine the selectivity of these inhibitors towards PP1 and PP2A. The synthetic strategy combined Enantio-switching and diastereo-switching method prompted us to synthesize a chemical probe which will clarify the selectivity of OKA and TTM towards PP. Towards this end, we have designed and synthesized a hybrid molecule between TTM and OKA, named as okadamycin **14** (*9*). This unnatural product inhibits PP2A more effectively than PP1, which supports our hypothesis that the portion of Segment C determined the selectivity of inhibitions to PP.

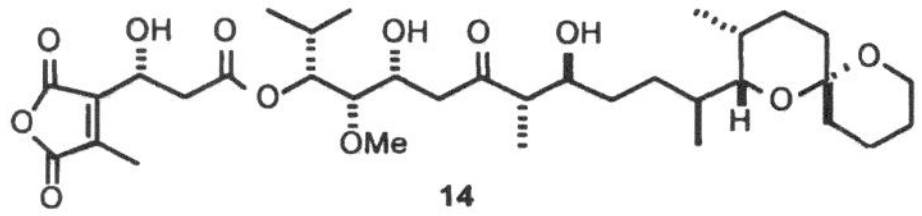

Figure 4A. Unnatural Inhibitor, okadamycin

New Unnatural Inhibitors of Protein Phosphatases, Heptanortautomycin

Further studies in this area proposed the second generation of such a hybrid molecule, namely heptanor-tautomycin **15**,as shown in Figure 4. This hybrid molecule has the structure of TTM with the enantiomeric form of Segment C of OKA **16**. We planned an improved route to **16** via **17**, which has been used in our total synthesis of TTM (*2*).

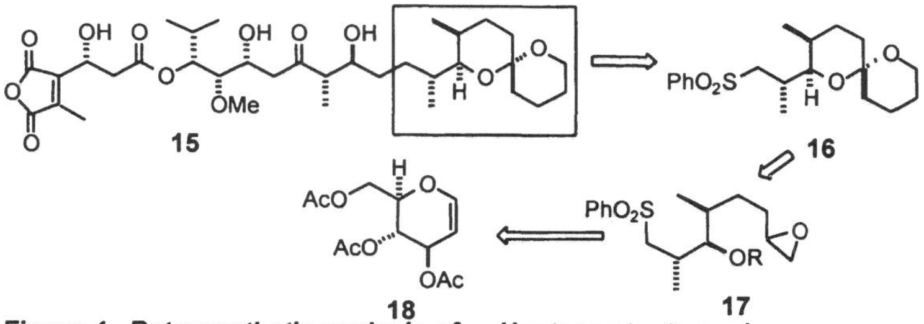

Figure 4. Retrosynthetic analysis of Heptanor-tautomycin

During previous synthesis of subsegment C1 of TTM **17** by using enantio-switching method, we have encountered a problem during epimerization of the intermediates as shown in Figure 5. The epimerization of**20** into **21** was difficult because of the axial methyl group in pyranose ring, which greatly diminished the driving force of thermodynamic equilibrium. In fact, the epimerization of the dicobalthexacarbonyl complex **20** with trifluoromethane sulfonic acid gave a mixture of **20** and its β-isomer **21** with the ratio of 1: 1.1, and the β-isomer **21** was separated by flash chromatography. After three recycling of the recovered α-isomer **20**, we obtained β-dicobalthexacarbonyl complex **21** in 65% yield. Decomplexation of **21** with iodine and hydrosilylation of **22** using 1mol% of sodium hexachloroplatinate (IV) gave **23** in 88% yield.

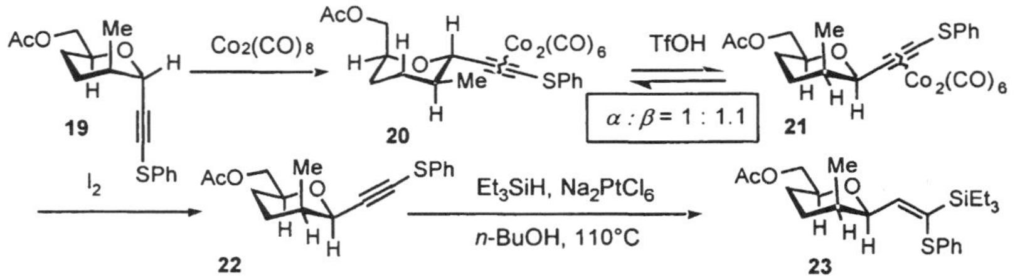

Figure 5. Previous synthtesis of Sub-segent C1 of TTM

To solve this problem, we modified the steps as show in Figure 6, where the methyl group be introduced after epimerization of the dicobalthexacarbonyl complex **25** which may give a 1: 4 mixture of **25** and **26** through our previous study (*10*). In addition, we have recently developed a new reductive decomplexation method of bis-cobalthexacarbonyl complex, which is applicable to the synthesis of **27** in a single step from bis-cobalthexacarbonyl complex **26** (*11*).

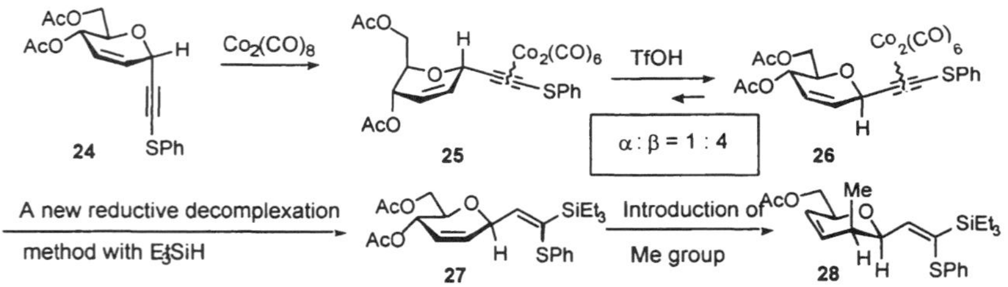

Figure 6. A new plan for the synthesis of Segement C1 of TTM

Synthesis of Heptanor-tautomycin

The synthetic plan in Figure 6 has been realized as full steps shown in Figure 7. *C*-Glycosidation of **18** with phenylthiotrimethylsilyl-acetylene and boron trifluoride etherate followed by treatment with biscobaltoctacarbonyl gave the biscobalthexacarbonyl complex **25** in 92% yield. Epimerization of the cobalt complex **25** was achieved with trifluoromethane sulfonic acid in dichloromethane at room temperature for 10 min under thermo-dynamically controlled condition.

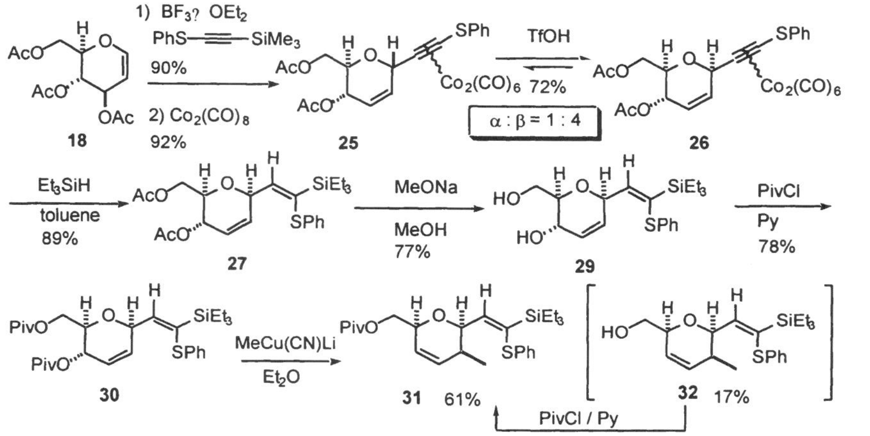

Figure 7. Epimerization and introduction of a methyl group

A 1: 4 mixture of **25** and **26** was obtained in 72% yield and the *β*-isomer **26** was separated by chromatography. Reductive decomplexation and hydrosilylation of **26** was achieved by treatment of **26** with triethylsilane in toluene at 65° C overnight to afford **27** in 89% yield. Introduction of the ring-methyl group was achieved by SN2' reaction of the allyl pivalate **30**, because hydrolysis of acetate was the major reaction in the case of **27**. Thus, treatment of **30** with lithium methyl cyanocuprate in ether at 0°C gave the methyl adduct **31** and **32** in 61and 17% yield respectively. Treatment of **32** with pivaloyl chloride gave **31** in 55% yield.

Synthesis of sub-segment C1 **17** and further transformation into enantiomeric Segment C of OKA **16** are shown in Figure 8. Heteroolefin **33** was prepared by oxidation of **31** with *m*-chloroperbenzoic acid, and heteroconjugate addition of methyllithium-lithium bromide complex to **33** was achieved by *α*-chelation controlled manner, and successive desilylation with tetrabutylammonium fluoride afforded **34** in 96% yield with high diastereoselectivity (*syn*: *anti* = >99: 1). Hydrogenation of the double bond of **34** in the presence of platinum oxide gave **35**. Treatment of **35** with methyltriphenoxyphosphorus iodide and reductive ring opening of the resulting **36** by zinc (*12*) furnished the open chain compound **37**. Protection of the alcohol **37** as *t*-butyldimethylsilyl ether followed by epoxidation of the olefin **38** with *m*-chloroperbenzoic acid furnished Sub-segment C-1 **17** as a mixture of two diastereomers. Epoxide ring opening by lithium acetylide in the presence of boron trifluoride diethyl etherate gave **39** in 96% yield. Hydrogenation of the triple bond of **39**, oxidation by PCC, removal of silyl ether protecting group and spiro-ketalization in refluxing methanol in the presence of *p*-toluenesulfonic acid completed the synthesis of **16** in 63% overall yield from **39**.

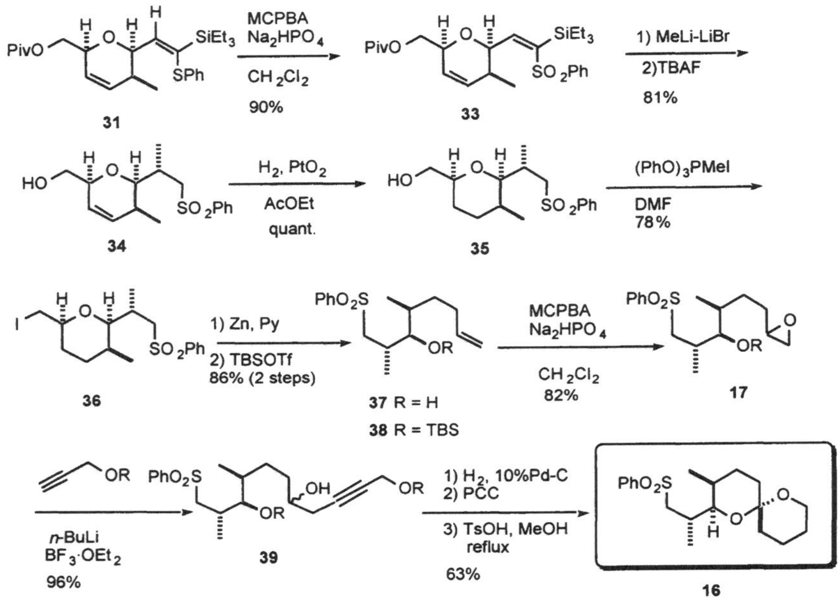

Figure 8. Synthesis of enantiomeric Segment C of OKA

With the efficient synthesis of **16** accomplished, the coupling reaction of Segment C and Segment B of TTM **40** (*13*) was undertaken. Treatment of **16** with *n*-butyllithium gave the corresponding sulfone carbanion, which reacted with **40** in the presence of borane trifluoride etherate to give **41.** Desulfonylation of **41** with sodium amalgam afforded **42** in 73% overall yield from **40.** Protecting group manipulation of **42** gave the diol **43** in 70%yield. Selective esterification of **43** with Segment A of TTM **44** under Yamaguchi condition and two-step deprotection involving removal of *t*-butyldimethylsilyl groups with poly (hydrogen fluoride) pyridine complex and cleavage of the two dithioketals using mercury perchlorate in aqueous acetonitrile furnished the synthetic Heptanor-tautomycin **15** in 52% overall yield from **43**.

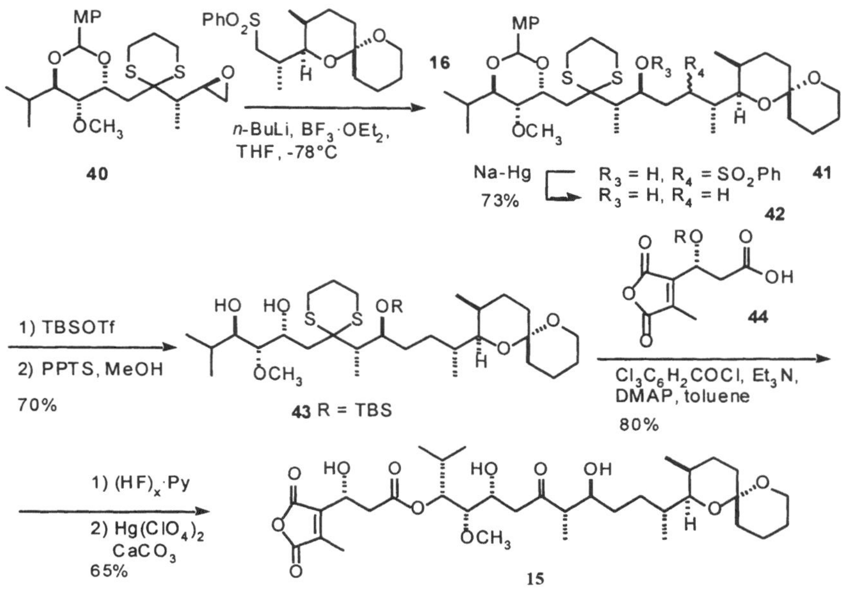

Figure 9. Synthesis of Heptanor-tautomycin

Synthetic Studies of Glycopeptide Mimics with Urea-glycosyl Bonds

In recent years, glycopeptides have become important area for bioorganic and medicinal research work because of its important biological activity. Research work in this area stimulated the development for the synthesis of glycopeptide mimics for medicinal studies and therapeutic applications. Glycopeptides have two modes for the attachment of glycosides to the peptide backbone involving either oxygen atom in the side chain of serine and threonine, or nitrogen atom in the side chain of asparagine. In the studies of glycopeptide mimics, *O*- and *N*-glycosyl linkages have been replaced by carbon-carbon, carbon-sulfur, and carbon-aminooxy bonds (*14*). In this project, we propose an approach to the synthesis of glycopeptide mimics in which *O*- and *N*-glycosyl linkages are replaced by urea-glycosyl bonds.

Retrosynthetic analysis for the synthesis of a key building block **I** for such a glycopeptide mimics is shown in Figure 10. It was envisaged that the synthesis of **I** could be achieved by a coupling reaction of the glycosyl isocyanates **II** and the amino acid derivatives **III**, and initial work focused on the synthesis of the glycosyl isocyanates IV.

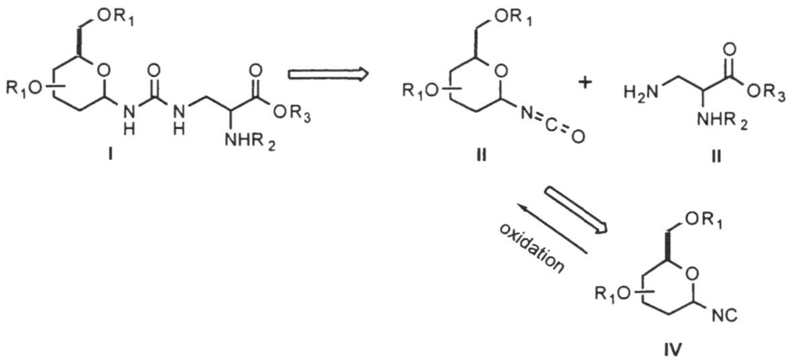

Figure 10. Retrosynthetic analysis for the glycopeptide mimics with urea-glycosyl bonds

The first synthesis of glycosylisocyanate has been reported in 1914 by E. Fisher, who described the reaction of teteraacetylbromoglucose with silver isocyante in xylene (*15*). Subsequent attempts to repeat his work by Johnson and Bergman found that two types of the glucosyl isocyantes A and B were formed, which suggested that this method seemed to suffer from non-stereospecificity (*16*). In this context, we have explored anew method for the stereospecific synthesis of the α-and β-glycosyl isocyanates and have now recognized hat oxidation of the glycosyl isocyanides **IV** is a good synthetic route for the preparation of the glycosyl isocyanates **II**. At the beginning stage of this route, the synthesis of the α- and β-glucosyl isocyanides has been developed, as shown in Figure 11.

Catalytic hydrogenation of the α-azide **45** afforded the α-glucosyl amine **46** which easily epimerized at the C-1 position to afford the thermodynamically stable β-glucosyl amine **47** due to the reverse anomeric effect (*17*). To avoid this problem, the reaction mixture was immediately treated with acetic formic anhydride after hydrogenation. This procedure afforded a mixture of the formamides **48** and **49** which was dehydrated under mild conditions (PPh_3, CBr_4, Et_3N) to give the glucosyl isocyanides **50** and **51** in the ratio of 81:19 (*18*) and the α-glucosyl isocyanide **50** was separated by silica-gel chromatography. The β-glucosyl isocyanide **51** was also prepared by a similar procedure starting from β-azide**52**. In this case, β-glucosyl amine **47** was stable, and easy convertible without epimerization at the C-1 position during the transformation from **47** to **51**.

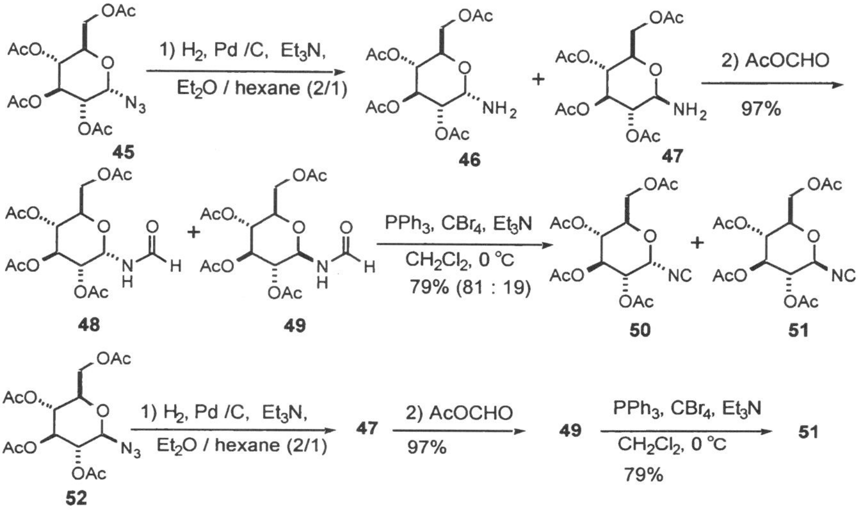

Figure 11. Synthesis of α- and β-glucosyl isocyanides

With both α- and β-glucosyl isocyanides in hand, we next examined oxidation of the glucosyl isocyanides. After extensive experimental efforts, we have finally realized that Method A (pyridine *N*-oxide in the presence of a catalytic amount of iodine) (*19*) and method B (2,4,6-trimethylbenzonitrileoxide) (*20*) are the most satisfactory oxidizing reagents. From the practical point of view, we preferred method A because pyridine *N*-oxide is commercially available. A typical example employing method A is shown in Figure 12. Oxidation of the α-glucosyl isocyanide **50** by method A in the presence of powdered molecular sieves 3A proceeded at room temperature for 30 min, and the resulting reaction mixture was immediately treated with cyclohexylamine to provide the α-glucosyl urea **54**. This reaction sequence involved the transformation of in situ-generated glucosyl isocyanate **53** into the stable glucosyl urea **54** in a one-pot process, which avoided the problem, associated with the isolation of the reactive glucosyl isocyanate **53**. The resulting glucosyl urea **54** was isolated in 91% yield after purification. A similar procedure starting from β-glucosyl isocyanide **51** gave β-glucosyl urea **56** in 95% yield. It should be noted that the epimerization at the anomeric position has never been observed during the transformation from the glucosyl isocyanides into the glucosyl ureas.

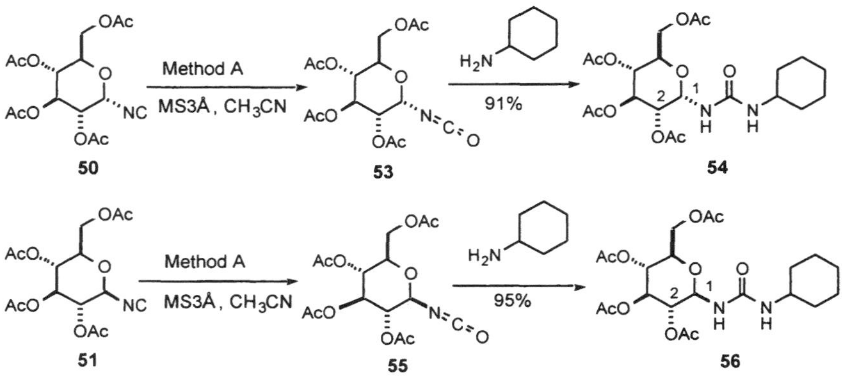

Figure 12. Stereospecific synthesis of theα- and β-glucosyl urea

The structures of the glucosyl urea **54** and **56** have been determined by NMR, that is, ^{13}C NMR analysis of the urea carbonyl carbon of **54** and **56** appeared 156.9 and 155.6 ppm, and ureido-glycosidic carbon appeared at 77.2and 80.2 ppm, respectively. A small coupling constant (5 Hz) of **54** between $H_1$and H_2, and large coupling constant (9 Hz) of **56** determined α- and β-stereochemistry at the anomeric positions.

Finally, the synthesis of a key building block for the glycopeptide mimics with urea-glycosyl bonds has been achieved, as shown in Figure 13.

Oxidation of **51** by method A and subsequent treatment of **55** with ammonium trifluoroacetate **57** in the presence of diisopropylethylamine provided **58** in 67% yield.

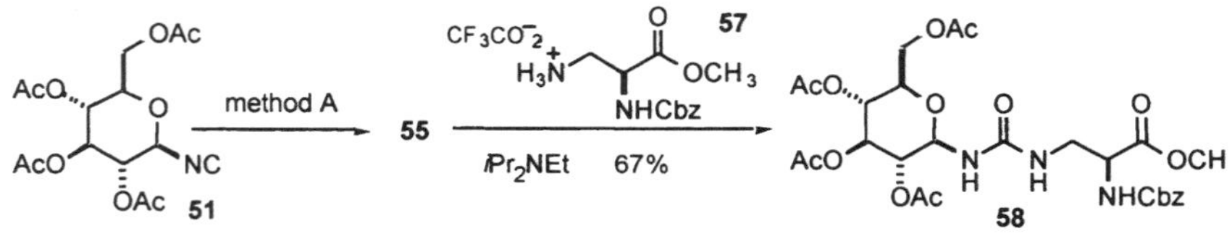

Figure 13. Synthesis of glycopeptide mimics with a urea-glycosyl

References

1. (a) Isobe, M.; Ichikawa, Y.; Masaki, H.; Goto, T. *Tetrahedron Lett.* **1984**, *25*, 3607. (b) Ichikawa, Y.; Isobe, M.; Masaki, H.; Kawai, T.; Goto, T. *Tetrahedron* **1987**, 43, 4759.
2. (a) Jiang, Y.; Ichikawa, Y.; Isobe, M. *Synlett.* **1995**, 285. (b) Jiang, Y.; Ichikawa, Y.; Isobe, M. *Tetrahedron* **1997,** 53, 5103.
3. (a) Nicholas, K. M.; Pettit, R. *Tetrahedron Lett.* **1971**, 347.
4. (a) Isobe, M.; Kitamura, M.; Goto, T. *Tetrahedron Lett.* **1979**, 3465; (b) Perspectives in the Organic Chemistry of Sulfur, ed. by B. Zwanenburg, A. J. H. Klunder, "New Synthetic Methods Using Vinyl Sulfones-Developments in Heteroconjugate Addition" M. Isobe, Studies in Organic Chemistry, 28, 1987; p 209, Elsevier Science Publishers.
5. (a) Tsukiyama, T.; Isobe, M. *Tetrahedron Lett.* **1992**, *33*, 7911: (b) M. Isobe, R. Nishizawa, S. Hosokawa, T. Nishikawa, J. *Chem. Soc., Chem. Commun.* **1998,** 2665.
6. (a) Isobe, M.; Jiang, Y. *Tetrahedron Lett.* **1995**, *36*, 567. (b) Jiang, Y.; Isobe, M. *Tetrahedron* **1996**, *36*, 2877.
7. Isobe, M.; Funabashi, Y.; Ichikawa, Y.; Mio, S.; Goto, T. *Tetrahedron Lett.* **1984,** 25, 2021.
8. (a) Sugiyama, Y.; Ohtani, I. I.; Isobe, M.; Takai, A.; Ubukata, M.; Isono, K. *Bioorg. Med. Chem. Lett.* **1996**, *6*, 3. (b) Takai, A.; Tsuboi, K.; Koyasu, K.; Isobe, M. *Biochem. J.* **2000**, *350*, 81.
9. Tsuboi, K.; Ichikawa, Y.; Isobe, M. *Synlett.* **1997,**713.
10. (a) Tanaka, S.; Tsukiyama,T.; Isobe, M.; *Tetrahedron Lett.***1993,** *34*, 5757; (b) Tanaka, S.; Isobe, M. *Tetrahedron* **1994,** *50*, 5633.
11. (a) Hosokawa, S.; Isobe, M. *Synlett.* 1995, 1179; b) Hosokawa, S.; Isobe, M. *Synlett.* **1996**, 351; c) Hosokawa, S.; Isobe, M. *J. Org. Chem.* **1999**, *65*, 37.
12. Bernet, B.; Vasella, A. *Helv. Chim. Acta.* **1979**, *62*, 1990.
13. Tsuboi, K.; Ichikawa, Y.; Jiang, Y.; Naganawa, A.; Isobe, M. *Tetrahedron,* **1997**, *53*, 5123.
14. Marcaurelle. L. A.; Bertozzi, C. R. *Chem. Eur. J.* **1999**, *5*, 1384.
15. Fisher, E. *Ber.* **1914**, *47*, 1377.
16. Johnson, T. B.; Bergmann, W. *J. Am. Chem. Soc.***1932**, *54*,3360.
17. Ogawa, T.; Nakabayashi, S.; Shibata, S. *Agric. Biol. Chem.* **1983**, *47 (29)*, 281.
18. (a) Nolte, R. J. M.; van Zomeren, A. J.; Zwikker, J. W. *J. Org. Chem.* **1978**, *43,* 1972. (b) Witczak, Z. J. *J. Carbohydr. Chem.* **1984**, *3 (3)*, 359.
19. Johnson, H. W.; Krutzsch, H. *J. Org. Chem.* **1967**, *32,* 1939.
20. Alpoim, C. M.; Barett, A.G. M.; Barton, D. H. R.; Hiberty, P. C. *Nouveau J. Chmie.* **1980**, *4*,127.

Indexes

Author Index

Subject Index

B

C

D

E

F

G

M

N

O

P

R

U

V

W

X